THEORY OF COMPUTATION SERIES

An Introduction to the General Theory of Algorithms

Michael Machtey
Paul Young

Purdue University

NORTH-HOLLAND·NEW YORK
NEW YORK • OXFORD • SHANNON

THE COMPUTER SCIENCE LIBRARY

Theory of Computation Series
PATRICK C. FISCHER, *Editor*

Borodin and Munro   The Computational Complexity
of Algebraic and Numeric Problems

Machtey and Young   An Introduction to the General Theory
of Algorithms

ELSEVIER NORTH-HOLLAND, INC.
52 Vanderbilt Avenue, New York, New York 10017

Distributors outside the United States and Canada:
THOMOND BOOKS
(A Division of Elsevier/North-Holland Scientific
        Publishers, Ltd.)
P.O. Box 85
Limerick, Ireland

Library of Congress Cataloging in Publication Data

Machtey, Michael.
  An introduction to the general theory of algorithms.

  (Theory of computation series) (Computer science library)
  Includes bibliographical references and index.
  1. Programming languages (Electronic computers)
2. Algorithms. 3. Recursive functions. I. Young,
Paul, 1936- joint author. II. Title.
QA76.7.M3   511'.8   77-22046
ISBN 0-444-00226-X
ISBN 0-444-00227-8 pbk.

Manufactured in the United States of America

Contents

Preface

In 1967, in the preface to his book *Computation: Finite and Infinite Machines*, Marvin Minsky, writing about the connections between computers and the general theory of algorithms said,

> The abstract theory . . . tells us in no uncertain terms that the machines' potential range is enormous, and that its theoretical limitations are of the subtlest and most elusive sort. There is no reason to suppose machines have any limitations not shared by man.
>
> We . . . expect our findings to crystallize in the years ahead—and we are confident that the ideas and the formal methods that fill this book will remain in the mainstream of that theory which is yet to come. We know this is so for several reasons: (a) The central concept of *effective computability* appears necessary, even inevitable, once one has grasped it . . . and (b) The shear simplicity of the theory's foundation and the extraordinarily short path from this foundation to its logical and surprising conclusions give the theory a mathematical beauty that alone guarantees it a permanent place in computer theory.

The decade that has passed since Minsky wrote these words has been a period of enormous growth and productivity, both for computer engineering and for computer science. There has been tremendous growth both in the sheer computational power which is available and in the dispersal of computational facilities to an evergrowing audience. Perhaps paradoxically, there has simultaneously emerged a much clearer, more precise understanding of the limits of computational power.

In the past ten years, Gödel's work on the undecidability of rich mathematical structures has been extended to a general method for showing that mathematical and computational structures, some of which have for years been known to be decidable, cannot possibly be decidable in any practical sense. Post's ideas for using reducibilities to classify the inherent difficulty of sets by their completeness properties have been applied to more concrete problems, leading to useful classifications of practical computational problems. Finally a whole theory of abstract complexity, applying to the most general imaginable programming systems and measures of complexity in terms of resource use has been developed, providing important new insights, such as the hopelessness of complete optimization in general purpose programming systems.

In 1967, the simplicity of the theory's foundations when formulated in sufficiently general terms was already largely apparent, but Minsky could only have guessed at much of the developing theory's later conclusions, mathematical beauty, and ability to demonstrate subtle theoretical limitations. Just as Minsky could write that the theory his book presented ten years ago was guaranteed a permanent place in computer theory, we are confident that much of the work of the past ten years is of such fundamental scientific importance that it too will occupy a permanent place in the theory of what is, and what is not, practically computable.

In this book we deal largely with absolutes, and as such the theory presented here will, in one form or another, endure through the years. It is only by understanding, in precise and well formulated ways, the *limits* on the computational method that one can hope to understand fully what *is* possible and hence to develop precise and well-formulated ideas about what is computationally feasible. The development of such precise tools should be a major goal of computer science, and it is for this reason that we believe that every computer scientist should understand both the underlying theory and the theoretical limitations presented here. We also hope that engineers, logicians, mathematicians, and philosophers interested in the computational process will find the theory interesting and enlightening.

This book is intended primarily as a text at the first or second graduate level. Mathematical maturity appropriate to graduate students in computer science is essential to mastery of the book, as is some minimal acquaintance with algorithms and programming. The speed with which one can cover the book, particularly Chapter 1, will depend on the students' background. Certainly some background in automata theory, formal language theory, or Turing machine theory is highly desirable. Since exercises are often an integral part of the material, the student will have to read the book with paper and pencil at hand, and an instructor may well want to incorporate many of the exercises into classroom presentations.

There is more than enough material here for a one semester course, and we anticipate most instructors omitting some of the optional material. We believe that by covering Chapters 1 through 3 and one additional chapter (with whatever appropriate definitions and basic results might be required from intermediate chapters), the book could be used for a one quarter course. For a full year course, an instructor would probably want to add supplementary material. The section on References and Further Readings should be helpful for this.

The book itself has developed from graduate and undergraduate courses which we have taught, at Purdue and elsewhere, for the past ten years. Half a generation of students have suffered through and contributed to the many revisions of the preparatory notes. We offer our apologies and thanks to them all. Paul Chew, Eric Dittert, Debby Joseph, and Tim Long have cheerfully assisted with the manuscript and galleys. Karl Winklmann's discussion of the material in Section 7.2 and some of the material in Section 2.3 has been particularly helpful. Our colleague Richard Buchi deserves special mention. Over the years his injunction always to keep in sight what is fundamental has been both a prodding thorn and an inspiration. Our debt to him can never be repaid. The work, the interest, and frequently the friendship of our colleagues around the world have also been an inspiration. Many, too numerous to mention individually, will see their own ideas, often recast in another form, presented here. We thank them all.

Michael Machtey

Summer, 1977   Paul Young

Introduction

This book is concerned with the general theory of algorithms and the programming languages used to implement them. We are interested in such questions as: What is an algorithm? What processes can or cannot be performed by algorithms? How can the languages used to implement algorithms be characterized? What is the computational complexity of an algorithm (program), and what is the nature of computational complexity? What processes can or cannot be performed by algorithms using a reasonable (practical) amount of computational resources? Since we are interested in the general theory of algorithms, we are concerned with getting mathematically precise answers to such questions with very widespread applications.

First, we would like to explain the title of this book, and at the same time, hopefully, add to the understanding of how the word *theory* should be used in computer science. According to the *Oxford English Dictionary*, theory comes from the Greek θεωρια, meaning a "looking at, viewing, contemplation, . . . ," and as one of its principal meanings, a theory is "that department of an art or technical subject which consists in the knowledge or statement of the facts on which it depends, or of its principles or methods, as distinguished from the *practice* of it." This is the sense in which we use the word theory in the title and throughout this book (and is the sense in which we believe it *should* be used in computer science). Notice that theory is distinguished from *practice*, not from *practical* or *useful*. In fact, it is hard to imagine the competent practice of computer science without an adequate understanding of the theory underlying the particular area of computer science being practiced, and in this sense, *all* theory is practical.

By an *algorithm* we mean a *recipe* or specific set of rules or directions for performing a task; in the context of this book, the word algorithm refers to recipes for performing tasks or solving problems which one might hope or expect to accomplish with a digital computer. By *general* we mean that we are primarily concerned with the general nature, behavior, complexity, etc. of algorithms, and not with the theory of specific algorithms or of algorithms for specific computational problems. When we deal with specific algorithms, it is usually for the purpose of developing the general theory of algorithms; when we deal with the theory of algorithms for specific computational problems, it is usually to illustrate principles or techniques from the general theory of algorithms. We have called this book "an introduction" because the core of the

1

book (as explained below) contains what we believe to be the essential minimum of the general theory of algorithms, and because we think this core is written so that it should be accessible to all computer scientists and mathematicians.

Although mathematicians have been interested in algorithms from the very beginning of mathematics, the study of the general *theory* of algorithms began fairly recently, arising in mathematical logic. Around 1900, logicians discovered several "paradoxes" in mathematics, particularly in set theory. These led to a great deal of work on the foundations of mathematics; parts of this work were concerned with "formal mathematical systems" and the intuitive notion of a "finite procedure." In the 1930's, several mathematical definitions of finite procedures were given by Church, Gödel, Herbrand, Kleene, Post, Turing, and others. Although these definitions of finite procedures were based on methods that appeared very different from each other (including mathematical closure operations, formal symbol manipulation systems, and idealized mechanical computers), all of the definitions turned out to be equivalent in the sense that they all "compute" the same class of functions. The class of functions given by these definitions is called the class of partial recursive functions. This class is important to computer science, and particularly to the general theory of algorithms, because all evidence indicates that the class of partial recursive functions is exactly the class of effectively computable functions; that is, that the partial recursive functions are exactly the functions which can be computed by finite procedures, algorithms, or computer programs. The evidence for this assertion is so strong that the assertion is given a special name, "the Church-Turing Thesis." In summary, the Church-Turing Thesis asserts that there is a nice mathematical characterization of the class of functions which are algorithmically computable.

With the general acceptance of the Church-Turing Thesis, the study of the computational processes used to define partial recursive functions was largely left behind. Beginning in the 1940's, recursive function theory studied the partial recursive functions as a mathematically given class of functions without regard to the particulars of *how* they are computed. This study had two basic aspects: the first related recursive functions to other branches of mathematics (for example, showing that there is no effective procedure for solving what are known as general diophantine equations), and the second consisted of extensive mathematical investigation of the recursive functions themselves. Because it deals with functions and sets *per se*, rather than with computations, very little of the latter work is of direct significance for computation.

After the growth of digital computers in the 1950's, a good deal of research began on the recursive functions in relation to computation, specifically on the nature of computational complexity. Originally, this work concerned specific models of computation such as Turing machines. Around 1965, Blum introduced a general (or "axiomatic") complexity theory which allowed the use of the full power of recursive function theory to derive very general results about computational complexity which hold for all models of computation, including those programming systems actually used in computing. Nevertheless, because such a general theory *cannot* possibly distinguish among different natural models of computation, and so far does not even distinguish natural models from unnatural models, work on specific models has continued in search of more specific information about actual computing. In particular, there has very recently been a great deal of work both on defining the "boundary" between computational problems which have practical algorithms for solving them and computational problems which, though algorithmically solvable, have no practical algorithms for solving them, and also on developing techniques for showing that particular computational problems (probably) do not have practical algorithms for solving them.

This book follows roughly the historical development outlined above, omitting most of the topics of purely mathematical interest. It is designed as a textbook for computer science students at the beginning graduate level; however, we believe that many other computer scientists and mathematicians will find it interesting and relevant. We assume very little specific background material in computer science, but some programming experience is very desirable, and well-developed intuitions about the nature of computing are indispensable. An undergraduate introduction to the general theory of computing, covering such topics as finite automata, formal grammars, and Turing machines, is also very desirable. Similarly, we assume very little specific mathematical background. The reader should be familiar with elementary set theory, as well as the basic concepts of what constitutes a mathematical proof, including proofs by mathematical induction. Both of these could be gained from, for example, either a good undergraduate computer science course in analysis of algorithms or discrete structures, or an undergraduate course in modern algebra.

The material in this book is presented in the order in which we feel that it is best encountered and digested. There are several optional sections and one optional chapter; these are identified in the Table of Contents. Subsequent nonoptional sections and chapters do not assume familiarity with this optional material, and can be read without having read the optional sections. The nonoptional sections are the core of the book, and they contain what might be called, "What every computer scientist should know about the general theory of algorithms." While optional Section 2.6 on the unsolvability of the Post Correspondence Problem is not an essential part of the general theory of algorithms, it develops an important tool very frequently used in many areas of computer science, including formal (programming) language theory; thus while that section is called optional in this book, it is part of "what every computer scientist should know." The remaining optional material deals with some very important and interesting applications of the theory of algorithms to mathematics, particularly to mathematical logic. Figure I.1 gives a pictorial representation of the division between the core sections and the optional sections of the book. We repeat that we have presented the material in this book in the order in which we feel that it is best encountered, and we *strongly recommend* that the reader who does not already have substantial familiarity with the general theory of algorithms not try either to jump ahead or to save optional material for the end.

Exercises in the body of the text that are not set off as "Additional

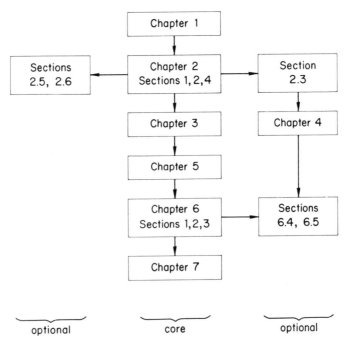

FIGURE I.1   The "logical" structure of the book.

Exercises'' at the end of a section are an integral part of the development of the material, and the reader should work them before proceeding; also, the solutions to many of these exercises are used later in the book. In most cases, these exercises are routine, and the reader who is not able to work one readily should immediately review the preceding material. In a few cases, these exercises may be slightly difficult; in such cases we indicate what we think the difficulties are, and we generally give a hint for working the problem when it is repeated at the end of the section (but the reader is cautioned to think about such exercises at least long enough to appreciate what the difficulties are before seeking the hint, and she is encouraged to make every effort to solve the exercise without the hint). In most sections, there is a set of additional exercises at the end of the section. Some of the additional exercises may also be essential to understanding the development of later material or may be assumed to have been solved later in the book; all such "required" exercises are routine and they are marked with an asterisk (whenever possible, we indicate for which later sections they are needed). All of the remaining additional exercises, while not essential to the development of the material, are intended to help deepen and extend the reader's understanding of the material, and we *strongly recommend* working as many of these as possible. Optional exercises range in difficulty from completely routine to very difficult, and we generally do not indicate what *we believe* the level of difficulty to be.

All items (propositions, definitions, figures, exercises, etc.) within a section are numbered *consecutively*, with the number indicating the chapter and section as well. Thus Exercise 3.1.4 may be found between Exercise 3.1.3 and Theorem 3.1.5, and all of these are in Section 3.1 of Chapter 3. We hope that this numbering scheme will help in locating such items when they are referred to elsewhere in the book.

We now conclude with an example of the sort of general property of algorithms we deal with in this book. Given any positive integer n it is trivial to write a program which prints n and halts. (We assume we are using a language like ALGOL or Turing machine programs in which constants can be arbitrarily large.) Consider the following *optimization problem*: for any integer n, find a shortest program which prints n and halts (the length of a program might be the number of characters in the program). It is clear that for each n there is in fact a shortest program which prints n (and there may well be several such shortest programs). However, we *can show* that there is no algorithm for finding such shortest programs; that is, there is no program **P** which on input n gives output **P**(n) which is such a shortest program. In establishing this *claim* we use the following three *assumptions* about our programming lan-

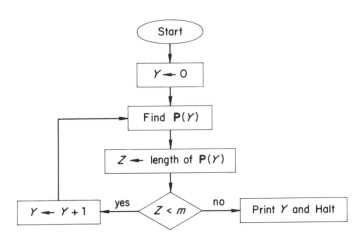

FIGURE I.2 Flowchart for the program Q_m.

guage: (1) given any program we can effectively find its length; (2) there are only finitely many programs of any given length; and (3) integer constants are represented in some base greater than or equal to 2. These assumptions are certainly very reasonable.

For the sake of a contradiction, assume **P** is a program which on input n gives output $P(n)$ which is a shortest program which prints n and halts. For each m, consider the program Q_m given by the flowchart in Figure I.2, which uses the program **P** as a subroutine. Given our assumption that **P** exists, then for each m, Q_m clearly can be programmed. What does Q_m do? It is not difficult to see that for each m, Q_m prints an integer i and halts if and only if no program of length less than m prints i and i is the smallest such integer; that is, Q_m prints the smallest integer i such that no program of length less than m prints i (and halts).

For each m, what is the length of Q_m? For different values of m, the only difference in the flow charts for the Q_m's is in the box that tests whether Z is less than m. Thus there is a constant c such that for all m,

$$\text{length of } Q_m = c + \text{length of ``}m\text{''},$$

The length of m (the space to write m as a constant) is no more than $1 + \log m$, and therefore

$$\text{length of } Q_m \leq c + 1 + \log m.$$

Thus for *all* sufficiently large values of m,

$$\text{length of } Q_m < m.$$

But then for any m large enough so that the length of $\mathbf{Q}_m$ is less than m, $\mathbf{Q}_m$ prints the smallest integer requiring a program of length at least m to print it, and $\mathbf{Q}_m$ *itself* has length less than m. This contradiction shows that our assumption that the program $\mathbf{P}$ exists must be false. Thus there is no program such as $\mathbf{P}$, and our claim is established.

The reader should compare the argument we have just given with Berry's Paradox: let k be "the least positive integer which is not denoted by any expression in the English language containing fewer than forty syllables." The English expression in quotes contains fewer than forty syllables and denotes the integer k.

Chapter 1

Models of Computation

In this chapter we develop a significant portion of the evidence for the fundamental Church-Turing Thesis, which asserts that the mathematically defined class of *partial recursive functions* is exactly the intuitively conceived class of effectively, that is algorithmically, computable functions. We give several different precise definitions of classes of intuitively computable functions, and we prove that all of these classes are the same in spite of the fact that they appear rather different. One of the definitions is based on mathematical closure operations, two are based on abstract versions of mechanical computers, and one is based on formal symbol manipulation systems. The strength of these proofs of equivalence as evidence for the Church-Turing Thesis is not just in the different appearance of the equivalent definitions. The nature of the proofs is such that we provide effective procedures, that is algorithms, for translating definitions (that is, "programs") in one system into equivalent definitions in any of the other systems. In addition, anyone familiar with the design of language translating programs, such as compilers, should have very little difficulty seeing that any of the modern programming languages can be translated in a similar fashion into one (and hence all) of our formulations, specifically into the random access machine (RAM) programs. Thus all of the functions which are effectively computable because there are programs for computing them are in the class of partial recursive functions.

1.1 WORDS OVER AN ALPHABET

We assume the reader is familiar with the usual mathematical vocabulary and notations dealing with sets and functions. In addition, we need some simple concepts and notations dealing with words over an alphabet. Let Σ be an arbitrary set of symbols. Σ^* stands for the set of all finite strings of symbols from Σ. The set Σ is called the *alphabet*, and the elements of Σ^* are called *words* over the alphabet Σ; in particular, Σ^* includes the *empty* (or *null*) *string* denoted by ϵ. We are interested in finite alphabets, and we use the notation A_n to stand for an alphabet $\{a_1, a_2, \ldots, a_n\}$ with n a positive integer.

In $A_3 = \{a_1, a_2, a_3\}$, if x is $a_2 a_3 a_1$ and y is $a_1 a_2$ then x concatenated with y is $a_2 a_3 a_1 a_1 a_2$ and it is denoted by xy. For any alphabet Σ, if x and y are in Σ^* then x *concatenated* with y is the string x followed by the

8

string y and is denoted by xy; x is called an *initial segment* of xy. The basic facts about concatenation are intuitively obvious, and we shall make no special effort to justify them. For example, each word x has a unique natural number *length* equal to the number of occurrences of alphabet symbols in it, and we denote the length of x by $|x|$; then for all strings x and y,

$$|xy| = |x| + |y|.$$

In addition, if X and Y are subsets of Σ^*, then

$$XY = \{xy: x \in X \text{ and } y \in Y\},$$
$$X^{(0)} = \{\epsilon\},$$
$$X^{(n+1)} = XX^{(n)} \quad \text{for } n \geq 0,$$

and

$$X^* = \bigcup_{n=0}^{\infty} X^{(n)}.$$

That is, XY is the set of all words gotten by concatenating a word from X with a word from Y, and $X^{(n)}$ is the set of all words gotten by concatenating together n words from X.

1.1.1 EXERCISE

(a) If all the words in X have length 1 (i.e., X is a set of alphabet symbols), show that $X^{(n)}$ consists of all words of length n using the alphabet X.

(b) If all words in X have length 3, what can you say about the lengths of words in $X^{(n)}$?

(c) If all words in X have even length, what can you say about the lengths of the words in $X^{(n)}$?

Notice that if X is a set of alphabet symbols then X itself is also an alphabet, and we have thus defined X^* both as the set of all words over X and as $\bigcup X^{(n)}$.

1.1.2 EXERCISE Let $X \subseteq \Sigma$; show that the two definitions given above for X^* both yield the same set.

We close this section by trying to clarify some possibly confusing notation. For any set K and any positive integer n, K^n is the n-fold Cartesian product of K with itself; that is, K^n is the set of all ordered n-tuples over the set K. If a is an alphabet symbol and n is a natural number, then a^n is the word consisting of n a's; note that for $n = 0$, $a^0 = \epsilon$. In light of the notation $X^{(n)}$ defined above, it would seem more natural to use $a^{(n)}$ instead of a^n to avoid confusion with the notation K^n;

however, a^n has become standard. Finally, N stands for the set of natural numbers (non-negative integers); that is, $N = \{0, 1, 2, 3, \ldots\}$. We make free use of the principle of mathematical induction to prove statements about the natural numbers, and we assume the reader is familiar with such proofs.

1.2 EFFECTIVELY COMPUTABLE FUNCTIONS

At a basic level, computation deals with the manipulation of symbols or characters. For example, the algorithm for addition taught in elementary school does not work directly on numbers, but works instead on the decimal representations of numbers. Thus we are concerned with computable functions which operate on symbols. In this section, before we define some specific models for computation, we look at some examples of functions to help sharpen our intuitive notion of effectively computable functions.

Consider the function $f : \{0, 1, \ldots, 9\}^* \to \{0, 1, \ldots, 9\}^*$ defined as follows:

$$f(x) = \begin{cases} 1 & \text{if} \quad \text{an } \textit{initial} \text{ run of digits in the decimal expansion of } \pi \\ & \qquad \text{is } x, \text{ ignoring the decimal point} \\ 0 & \text{otherwise.} \end{cases}$$

For example, $f(1) = 0$, $f(314) = 1$, $f(325) = 0$, etc. This function is effectively computable.

1.2.1 EXERCISE   Give an algorithm for computing f.

Next, consider the function $g : \{0, 1, \ldots, 9\}^* \to \{0, 1, \ldots, 9\}^*$ defined as follows:

$$g(x) = \begin{cases} 1 & \text{if} \quad \textit{some} \text{ run of digits in the decimal expansion} \\ & \qquad \text{of } \pi \text{ is } x, \text{ ignoring the decimal point} \\ 0 & \text{otherwise.} \end{cases}$$

For example, $g(1) = 1$, $g(14) = 1$, $g(31) = 1$, and $g(71895) = ?$. This function is *probably* not effectively computable; we would certainly have to know a great deal more about the decimal expansion of π than we do now to compute g.

1.2.2 EXERCISE   Try to give an algorithm for computing g, and think about your difficulties.

As a final example, consider $h : \{0, 1, \ldots, 9\}^* \to \{0, 1, \ldots, 9\}^*$

defined as follows:

$$h(x) = \begin{cases} 1 & \text{if} \quad \text{there is a run of } at \text{ } least \text{ } x \text{ consecutive} \\ & \qquad 7\text{'s in the decimal expansion of } \pi \\ 0 & \text{otherwise,} \end{cases}$$

where we view x as an integer by ignoring any initial 0's. You may find it surprising that we can easily show that this function *is* effectively computable, even though we may be at a loss to say which of an infinity of algorithms is the correct one for computing it. This is true because either there are arbitrarily long runs of consecutive 7's in the decimal expansion of π or else there is a longest run with, say, m 7's. In the first case, $h(x) = 1$ for all x, while in the second case there is *some* integer m such that $h(x) = 1$ if $x \le m$ and $h(x) = 0$ if $x > m$. In any case, there is a simple algorithm for computing h, even though we do not *know* what that particular algorithm happens to be.

1.3   THE PRIMITIVE RECURSIVE FUNCTIONS

We now begin developing our first model of computation. Our first goal in developing such models is to provide a precise mathematical characterization of the class of all computable functions. Our first model will seem initially to have more the flavor of mathematics than of computer science. We begin by fixing an alphabet A_k and defining a set of functions from $(A_k{}^*)^n$ into $A_k{}^*$ (for $n > 0$) which are clearly, in our intuitive sense, effectively computable. We start by using three definition schemes to define infintely many *base* functions.

Scheme I   The *erase* (or *null*) function on any x in $A_k{}^*$ is given by

$$E(x) = \epsilon.$$

Given any word x, the function E simply erases it.

Scheme II   For each j such that $1 \le j \le k$, the jth *successor* function on any x in $A_k{}^*$ is

$$S_j(x) = xa_j.$$

Given any word x, the function S_j simply concatenates a_j to the right end of x.

Scheme III   For each n and j such that $1 \le j \le n$, the *projection* function $P_j{}^n$ on any $(x_1, \ldots, x_n)$ in $(A_k{}^*)^n$ is

$$P_j{}^n(x_1, \ldots, x_j, \ldots, x_n) = x_j.$$

Given any list of n words, $P_j{}^n$ simply picks out the jth word in the list.

Clearly each of the functions defined above is effectively computable, and certainly not every computable function is among those defined above. Notice that while Schemes I and II give only finitely many functions, Scheme III gives infinitely many different functions and so is necessarily a scheme rather than a single explicit definition.

1.3.1 EXERCISE   What is a common name for the function $P_1{}^1$?

For each $n \geq 2$, the erase function E^n of n arguments defined by $E^n(x_1, \ldots, x_n) = \epsilon$ is not among the base functions defined in Schemes I, II, and III, even though it is certainly computable.

1.3.2 EXERCISE   What rules for combining functions would enable us to obtain E^n *easily* from the base functions defined above?

Obviously there are *many* other computable functions which are not among the base functions. As mentioned above, our goal in this chapter is to characterize the class of *all* computable functions, and a remarkable fact is that we can actually obtain all computable functions by starting with these simple base functions and combining them in a few straightforward ways. We begin by giving two operations for combining functions.

Scheme IV   For positive integers m and n let g be a given function of m arguments, and let $h_1, h_2, \ldots, h_m$ be given functions each of n arguments. The function f of n arguments is obtained from g and $h_1, \ldots, h_m$ by *substitution* if, for all $x_1, \ldots, x_n$ in $A_k{}^*$,

$$f(x_1, \ldots, x_n) = g(h_1(x_1, \ldots, x_n), \ldots, h_m(x_1, \ldots, x_n)).$$

This operation is a generalization of composition of functions, and in fact it is often referred to as *composition*.

Scheme V   For any positive integer n let g be a given function of $n - 1$ arguments, and let $h_1, h_2, \ldots, h_k$ be given functions of $n + 1$ arguments, all over $A_k{}^*$. The function f of n arguments is obtained from g and $h_1, \ldots, h_k$ by *(primitive) recursion* if, for all y and $x_2, \ldots, x_n$ in $A_k{}^*$,

$$f(\epsilon, x_2, \ldots, x_n) = g(x_2, \ldots, x_n)$$

and

$$f(ya_1, x_2, \ldots, x_n) = h_1(y, f(y, x_2, \ldots, x_n), x_2, \ldots, x_n)$$
$$f(ya_2, x_2, \ldots, x_n) = h_2(y, f(y, x_2, \ldots, x_n), x_2, \ldots, x_n)$$
$$\vdots$$
$$f(ya_k, x_2, \ldots, x_n) = h_k(y, f(y, x_2, \ldots, x_n), x_2, \ldots, x_n).$$

Notice that if $n = 1$, this definition does not really make sense; what we mean in this case is that a function f of one argument is obtained by recursion from the "constant" z in $A_k{}^*$ and given functions $h_1, \ldots, h_k$ of two arguments if $f(\epsilon) = z$ and for all y in $A_k{}^*$, $f(ya_i) = h_i(y, f(y))$ for $1 \leq i \leq k$. (Notice that we think of a function of zero arguments as a constant.)

The operation of primitive recursion gives a limited type of recursive definition of functions; the value of f with ya_i as its first argument is allowed to depend on the value of f with y as its first argument, as well as on the other arguments. Notice that having the (possibly) different functions $h_1, \ldots, h_k$ allows a primitive recursive definition to perform a "test" to find the last letter in the first argument. For example, suppose we are working over $A_2 = \{a_1, a_2\}$ and for all x and y, $h_1(x, y) = ya_1$ and $h_2(x, y) = y$; then if f is obtained by recursion from ϵ, h_1, and h_2 the reader can verify that for all x, $f(x)$ is x with all occurrences of a_2 removed.

1.3.3 EXERCISE Explain how to use algorithms for computing the given functions g and the h_i's as subroutines in algorithms for computing the functions f obtained from them by substitution and recursion. (The reader familiar with high level programming languages such as ALGOL-60 should be able to work this exercise in two ways: one using recursive procedure definitions and the other using a **for**-loop construction.)

A function f mapping $(A_k{}^*)^n$ into $A_k{}^*$ is *primitive recursive* (over $A_k{}^*$) if f can be obtained from the base functions by (finitely many) applications of the operations of substitution and recursion. That is, the class of primitive recursive functions is the smallest class of functions which contains the base functions and which is closed under the operations of substitution and primitive recursion. Note that by the exercise above, there is an algorithm to compute any given primitive recursive function. In fact, from the exercise you should see that such an algorithm can easily be obtained *from the definition* of the function via substitution and recursion from the base functions. We can easily think of the definition of a primitive recursive function as a sort of program for computing the function. Specifically, we can define the class of *primitive recursive programs* for the alphabet A_k as follows: **E** is a program of one argument; for each j $(1 \leq j \leq k)$, S_j is a program of one argument; for each i and n $(1 \leq i \leq n)$, $P_i{}^n$ is a program of n arguments; if **H** is a program of m arguments and $G_1, \ldots, G_m$ are each programs of n arguments, then $H(G_1, \ldots, G_m)$ is a program of n arguments; finally, if either **G** is a word in $A_k{}^*$ and each H_j $(1 \leq j \leq k)$ is a program of two arguments or if **G** is a program of n arguments and each H_j $(1 \leq j \leq k)$ is a program of $n + 2$ arguments, then $PR(G, H_1, \ldots, H_k)$ is a program of $n + 1$ arguments. The preceding sentence defines the *syntax*, or format, for the primitive recursive programs; their *semantics*, or meaning, is given by the functions which the programs compute. Obviously, the programs **E**, S_j, $P_j{}^n$ are intended to compute the base functions. If for any program **P** of n arguments we denote the function it computes by $P(x_1, \ldots, x_n)$, then the function computed by $H(G_1, \ldots, G_m)$ is defined by the equation

$$H(G_1, \ldots, G_m)(x_1, \ldots, x_n)$$
$$= H(G_1(x_1, \ldots, x_n), \ldots, G_m(x_1, \ldots, x_n))$$

and $PR(G, H_1, \ldots, H_k)(x_1, \ldots, x_{n+1})$ is defined by the $k + 1$ equations

$$PR(G, H_1, \ldots, H_k)(\epsilon, x_2, \ldots, x_{n+1}) = G(x_2, \ldots, x_{n+1})$$

and

$$PR(G, H_1, \ldots, H_k)(za_i, x_2, \ldots, x_{n+1})$$
$$= H_i(z, PR(G, H_1, \ldots, H_k)(z, x_2, \ldots, x_{n+1}), x_2, \ldots, x_{n+1})$$

A complete proof that a function f is primitive recursive would be a proof that some particular primitive recursive program does indeed compute f. Of course in practice, if we wish to prove that some function is primitive recursive we do not always need to give a complete definition of it from the base functions using substitution and recursion; it is sufficient to give a definition of the function via substitution and recursion from other functions we already know to be primitive recursive.

To illustrate how functions are proved to be primitive recursive, and also to begin building a stockpile of primitive recursive functions which will be useful to us later, we next define some primitive recursive functions. For each fixed word $x = a_{i_1} a_{i_2} \ldots a_{i_n}$ in $A_k{}^*$ the *constant function* f_x such that $f_x(y) = x$ for all y is primitive recursive, and it is defined from the erase function and the successor functions by n

applications of substitution as follows:

$$f_x(y) = S_{i_n}(\ldots (S_{i_2}(S_{i_1}(E(y)))) \ldots).$$

Actually, we have been a bit sloppy here. We are asserting that infinitely many functions are all primitive recursive, and strictly speaking we should prove by induction on the length of x that each f_x is primitive recursive. Such an inductive proof would observe that $f_\epsilon = E$, the erase function, and that for all x and $1 \leq i \leq k$,

$$f_{xa_i}(y) = S_i(f_x(y))$$

for all y. However, having once pointed this out, we shall henceforth generally be sloppy and not write out such simple inductions explicitly. We caution the reader not to confuse such inductions dealing with infinitely many functions with the definition of a single function by recursion; even though the notations look similar, the conceptual difference is quite important and it is not proper to replace a definition by recursion by some convenient notation involving "$\ldots$".

For each $n > 1$, the *concatenation function* con_n of n arguments such that $con_n(x_1, x_2, \ldots, x_n) = x_1 x_2 \ldots x_n$ for all $x_1, \ldots, x_n$ is primitive recursive. *If* we have that con_n is primitive recursive, *then* it is easy to show that con_{n+1} is primitive recursive since

$$con_{n+1}(x_1, \ldots, x_n, x_{n+1}) = con_2(con_n(x_1, \ldots, x_n), x_{n+1}).$$

But notice that this is not a correct substitution according to Scheme IV; the correct version is

$$con_{n+1}(x_1, \ldots, x_{n+1}) = con_2(con_n(P_1^{n+1}(x_1, \ldots, x_{n+1}), \ldots,$$
$$P_n^{n+1}(x_1, \ldots, x_{n+1})), P_{n+1}^{n+1}(x_1, \ldots, x_{n+1})).$$

1.3.4 EXERCISE Make sure you understand the distinction we have just made.

Thus we are left with the task of showing that con_2 is primitive recursive.

1.3.5 EXERCISE Show that con_2 is primitive recursive.

This exercise is a little more difficult than you may realize at first. In working problems like this you must be careful not to use your knowledge that the function is easily computable; instead, you should remember that definitions of primitive recursive functions are in a programming language with a very restrictive syntax (format), and force yourself to stay within it. For example

$$con_2(x_1, \epsilon) = P_1^1(x_1)$$
$$con_2(x_1, ya_i) = S_i(P_2^3(y, con_2(x_1, y), x_1))$$

is not a legitimate primitive recursive definition of con_2 since according to Scheme V primitive recursion can only use recursion on the *first* argument. We give a hint of how you can circumvent this difficulty in the restatement of Exercise 1.3.5 at the end of this section; however, you should make every effort to work the exercise before consulting the hint. When you see how to work this exercise you will know how to justify primitive recursions on any *one* argument. The reason for using such a restricted language is that although programming (that is, defining functions) in such a restricted language is tedious, the simplicity of the language greatly facilitates proofs *about* the language.

We continue now with showing that some more simple functions are primitive recursive. For each $n > 1$, the function $[\ ]^n$ such that $[x]^n = x^n$, the n-fold concatenation of x with itself, is primitive recursive; although $[x]^n = con_n(x, \ldots, x)$ is *not* a valid primitive recursive definition, $[x]^n = con_n(P_1^{\ 1}(x), \ldots, P_1^{\ 1}(x))$ is valid. The *delete last* function *dell* such that

$$dell(x) = \begin{cases} \epsilon & \text{if } x = \epsilon \\ x \text{ with its last letter deleted} & \text{if } x \neq \epsilon \end{cases}$$

is primitive recursive, with definition

$$dell(\epsilon) = \epsilon \quad \text{and} \quad dell(xa_i) = P_1^2(x, dell(x))$$

for all $1 \leq i \leq k$; notice that in this definition it happens to be the case that all of the functions $h_1, \ldots, h_k$ are the same. Also the *level* function *lev* such that

$$lev(x) = \begin{cases} \epsilon & \text{if } x = \epsilon \\ a_1 & \text{if } x \neq \epsilon \end{cases}$$

has primitive recursive definition

$$lev(\epsilon) = \epsilon \quad \text{and} \quad lev(xa_i) = f_{a_1}(P_1^2(x, lev(x))).$$

1.3.6 EXERCISE Why did we not simply define $lev(xa_i) = f_{a_1}(x)$?

Now that we have given these examples of detailed primitive recursive definitions, we shall henceforth allow ourselves the freedom of defining primitive recursive functions by recursion on *any one* of the arguments and in general committing any one of the minor transgressions we have just discussed *because* we now know how to remedy such failures and put the definitions in the correct format. You should allow yourself the same freedom *after* you have gotten some practice in writing detailed primitive recursive definitions by correctly working the next exercise.

1.3.7 EXERCISE Show, in complete detail, that each of the following functions is primitive recursive:

(a) the function *lev'* such that

$$lev'(x) = \begin{cases} a_1 & \text{if } x = \epsilon \\ \epsilon & \text{if } x \neq \epsilon \end{cases}$$

(b) for each j such that $1 \leq j \leq k$, the function end_j such that

$$end_j(x) = \begin{cases} a_1 & \text{if } x \text{ ends in } a_j \\ \epsilon & \text{otherwise} \end{cases}$$

(c) the *reverse* function *rev* such that $rev(x)$ is x reversed
(d) the *delete first* function *delf* such that

$$delf(x) = \begin{cases} \epsilon & \text{if } x = \epsilon \\ x \text{ with its first letter deleted} & \text{if } x \neq \epsilon \end{cases}$$

(e) the *minus* function "$-$" such that $x - y$ is x minus its first $|y|$ letters

Additional Exercises

*1.3.5 Show that con_2 is primitive recursive: first show that con' such that $con'(x, y) = yx$ is primitive recursive, then define con_2 from con' using projection functions to "switch" the arguments. Note that by this method we can give legitimate primitive recursive definitions of functions defined by recursion on *any single* argument.

1.3.8

(a) Show that our set of base functions is not minimal by defining P_1^n from the erase function, the successor functions, P_1^{n-1}, and P_2^{n+1} by substitution and recursion, for any $n > 1$. Thus we could afford to throw away infinitely many of the base functions and we would still get all of the primitive recursive functions.
(b) Show that on the other hand, infinitely many base functions are necessary, by showing that not all of the primitive recursive functions can be obtained by substitution and recursion from any finite subset of the base functions.

1.4 SOME HELPFUL PRIMITIVE RECURSIVE FUNCTIONS AND PREDICATES

In this section we enlarge our arsenal of primitive recursive functions for later use. Although it is possible to think of all algorithms as computing functions by taking a list of words as input data and returning some word as output, there are many occasions when this way of thinking is very unnatural. Sometimes it makes more sense to think of an algorithm as testing some property of its inputs and returning one of the Boolean values "true" or "false," or to think of an algorithm as taking some variables of type Boolean as input and returning a string as output. For example, the functions end_j in Exercise 1.3.7 essentially test whether the input word x ends in the letter a_j. One possible method for dealing with properties would be to expand the basic definition schemes in the previous section, but this would result in much more complicated proofs *about* that definition. Instead, we choose to remain formally restricted to functions and to develop mechanisms which will allow us to deal with such things as properties and Boolean variables within the formal context of functions, but in informal ways which are much more appropriate to our intuition.

Modern mathematics has a very simple mechanism for dealing with properties. An (n-*ary*) *predicate* (*on* A_k^*) is simply a subset of $(A_k^*)^n$. Thus the binary "property" of x and y being the same word is simply the predicate $\{(x, y): x = y\}$. If P is an n-ary predicate we usually write $P(x_1, \ldots, x_n)$ instead of $(x_1, \ldots, x_n) \in P$, with the intuitive meaning that the list $x_1, \ldots, x_n$ has the "property" P; note that the intuitive content of $x_1, \ldots, x_n$ having the "property" P may in general be no more than that of the n-tuple $(x_1, \ldots, x_n)$ being a member of the set P, although specific predicates we deal with almost always have a reasonable intuitive meaning or "property." In particular, if P is a binary predicate we often write xPy instead of $P(x, y)$ or $(x, y) \in P$; thus when (x, y) is in the equality predicate we write $x = y$, which is only natural.

Mathematicians also have a very simple mechanism for identifying predicates with functions. The *characteristic function* of a predicate P is the function c_P such that for all $x_1, \ldots, x_n$

$$c_P(x_1, \ldots, x_n) = \begin{cases} a_1 & \text{if } (x_1, \ldots, x_n) \in P \\ \epsilon & \text{if } (x_1, \ldots, x_n) \notin P. \end{cases}$$

We can now easily expand our notion of primitive recursive "computability" to include predicates by defining a predicate P to be *primitive recursive* if c_P is primitive recursive. Thus the function *lev* from the last section is the characteristic function of the primitive recursive property of being a word different from ϵ.

There are some fundamental things we would like to be able to do with predicates, such as combining them by taking Boolean combinations. To this end, if P and Q are n-ary predicates we define *notP* to be the complement of P, *PorQ* to be $P \cup Q$, and *PandQ* to be $P \cap Q$; a moment's reflection should convince you that these definitions are intuitively correct. The next proposition is the first of several showing how we can handle predicates in intuitively comfortable ways within the restrictions of primitive recursive "computability."

1.4.1 PROPOSITION *If P and Q are primitive recursive predicates then so are notP, PorQ, and PandQ.*

Proof

$$c_{notP}(x_1, \ldots, x_n) = lev'(c_P(x_1, \ldots, x_n));$$
$$c_{PorQ}(x_1, \ldots, x_n) = lev(con_2(c_P(x_1, \ldots, x_n), c_Q(x_1, \ldots, x_n)));$$
and
$$c_{PandQ}(x_1, \ldots, x_n) = dell(con_2(c_P(x_1, \ldots, x_n), c_Q(x_1, \ldots, x_n))). \qquad \square$$

For later purposes it is important to note that this proof, and the proofs of the two propositions which follow, actually prove that *any* class of functions which includes the primitive recursive functions and is closed under substitution and primitive recursion then has its predicates (characteristic functions) closed under the operations indicated in the proposition. The next proposition justifies using the familiar way of defining a function by cases within the primitive recursive formalism.

1.4.2 PROPOSITION (Definition by Cases) *If $P_1, \ldots, P_n$ are pairwise disjoint primitive recursive predicates and $f_1, \ldots, f_n, f_{n+1}$ are primitive recursive functions, then the function g defined below is also primitive recursive:*

$$g(x_1, \ldots, x_m) = \begin{cases} f_1(x_1, \ldots, x_m) & \text{if} \quad P_1(x_1, \ldots, x_m) \\ \quad \vdots & \quad \vdots \qquad \vdots \\ f_n(x_1, \ldots, x_m) & \text{if} \quad P_n(x_1, \ldots, x_m) \\ f_{n+1}(x_1, \ldots, x_m) & \text{otherwise.} \end{cases}$$

Proof First notice that since $P_1, \ldots, P_n$ are pairwise disjoint, for each $(x_1, \ldots, x_m)$ exactly one case in the specification of g applies, and thus the specification of g makes sense. For the sake of simplifying the notation, we only do the case where $m = 1$ and $n = 2$. First we define the handy function $\#: (A_k{}^*)^2 \rightarrow A_k{}^*$ such that

$$x \# y = \begin{cases} \epsilon & \text{if} \quad x = \epsilon \\ y & \text{if} \quad x \neq \epsilon \end{cases}$$

and we observe that $\#$ is primitive recursive since $\epsilon \# y = E(y)$ and $xa_i \# y = P_3{}^3(x, x \# y, y)$. But then

$$g(x) = con_3(c_{P_1}(x) \# f_1(x), c_{P_2}(x) \# f_2(x), c_{not(P_1 or P_2)}(x) \# f_3(x));$$

from this example it should be clear how the proof would look for any m and n. □

Next, we define the application of "bounded quantifiers" to predicates.

1.4.3 DEFINITION (Bounded Quantifiers)   If P is a predicate with $n+1$ arguments $y, z_1, \ldots, z_n$, then

$\exists y/xP(y, z_1, \ldots, z_n)$ if and only if   there is *some* initial segment
$\qquad\qquad\qquad\qquad\qquad\qquad\qquad\qquad$ y of x such that $P(y, z_1, \ldots, z_n)$

and

$\forall y/xP(y, z_1, \ldots, z_n)$ if and only if   for *all* initial segments
$\qquad\qquad\qquad\qquad\qquad\qquad\qquad\qquad$ y of x, $P(y, z_1, \ldots, z_n)$.

(Notice that if P is a predicate with the $n+1$ arguments $y, z_1, \ldots, z_n$, then the truth or falsity of $\exists y/xP(y, z_1, \ldots, z_n)$ and of $\forall y/xP(y, z_1, \ldots, z_n)$ depends on x and on $z_1, \ldots, z_n$, but not directly on y. Thus the arguments for $\exists y/xP(y, z_1, \ldots, z_n)$ and for $\forall y/xP(y, z_1, \ldots, z_n)$ are x and $z_1, \ldots, z_n$. We frequently abbreviate $\exists y/xP(y, z_1, \ldots, z_n)$ to $\exists y/xP$ and similarly for $\forall y/xP$.)

1.4.4 PROPOSITION   *If P is a primitive recursive predicate, then $\exists y/xP$ and $\forall y/xP$ are also primitive recursive predicates.*

Proof   Let $c_{\exists/P}$ be the characteristic function of the predicate $\exists y/xP$. Then we can define $c_{\exists/P}$ by recursion as follows:

$$c_{\exists/P}(\epsilon, z_1, \ldots, z_n) = c_P(\epsilon, z_1, \ldots, z_n)$$

and

$$c_{\exists/P}(xa_i, z_1, \ldots, z_n) = lev(con_2(c_{\exists/P}(x, z_1, \ldots, z_n), c_P(xa_i, z_1, \ldots, z_n))).$$

When you figure out what these equations are saying you will see that they simply assert that

$$\exists y/\epsilon P(y, z_1, \ldots, z_n) \qquad \text{iff} \qquad P(\epsilon, z_1, \ldots, z_n)$$

and

$$\exists y/xa_iP(y, z_1, \ldots, z_n) \quad \text{iff} \quad \exists y/xP(y, z_1, \ldots, z_n) \text{ or } P(xa_i, z_1, \ldots, z_n)$$

which is certainly much more readable. Since the second version is simply a *notational* variation of the first, in future definitions and proofs we use the second form and abandon the first, even though it is the first which is formally correct. Finally,

$$\forall y/xP(y, z_1, \ldots, z_n) \quad \text{iff} \quad not \exists y/x \, notP(y, z_1, \ldots, z_n). \quad \Box$$

As an illustration of the use of the previous three propositions, we now show that the equality predicate is primitive recursive. The predicate $x = \epsilon$ is primitive recursive since lev' is its characteristic function. The predicate $end(x) = end(y)$ which holds if and only if the last letters in x and y are the same is primitive recursive because

$$end(\epsilon) = end(y) \quad \text{iff} \quad y = \epsilon$$

and

$$end(xa_i) = end(y) \quad \text{iff} \quad end_i(y).$$

1.4.5 EXERCISE Rewrite the recursive definition of the predicate $end(x) = end(y)$ using the characteristic function notation as in the proof of the previous proposition.

The predicate $|x| = |y|$ is primitive recursive since $|x| = |y|$ iff $x - y = \epsilon$ and $y - x = \epsilon$. Now we can show that the equality predicate is primitive recursive:

$$x = y \quad \text{iff} \quad |x| = |y| \text{ and } \forall z/x[end(z) = end(rev(rev(y) - (x - z)))],$$

which can be seen to assert that x is equal to y if and only if x and y have the same length and every initial segment of x ends in the same symbol as the corresponding initial segment of y.

As our last project in this section we show how to use the natural numbers (via the lengths of words) to code all the words in A_k^* in such a way that the coding and decoding functions are primitive recursive. To this end, we first observe that we can easily regard words in A_1^* as integers by identifying a_1^n with the natural number n. We can then code A_k^* onto A_1^* by thinking of words in A_k^* as integers written to the "base k," but in a way that may be slightly different from what you are used to. Since we want a one-to-one correspondence, we do not use the standard coding where $0031 = 31$. We let ϵ represent zero and the string $a_{i_1} \cdots a_{i_n}$ represent $\Sigma_{1 \leq j \leq n} \, i_j k^{n-j}$. Thus for $k=10$, $a_1, \ldots, a_{10}$

represent the "digits" $1, \ldots, 10$ instead of $0, \ldots, 9$, and so $a_3 a_1$
represents 31 and $a_{10} a_1$ represents 10 "tens" and 1 "one" which is 101;
isn't that amazing? But, $a_{10} a_{10}$ represents 110 and $a_1 a_1 a_1$ represents 111,
so we cannot always simply read off the subscripts. This example, along
with a little reflection, should convince you that our representation gives
a one-to-one correspondence between A_k^* and N. Thus the coding
function C_k from A_k^* to A_1^* defined by $C_k(\epsilon) = \epsilon$ and

$$C_k(a_{i_1} \ldots a_{i_n}) = a_1^{\sum_{1 \le j \le n} i_j k^{n-j}}$$

is easily seen to be one-to-one and onto. Moreover, C_k is primitive
recursive since $C_k(\epsilon) = \epsilon$ and $C_k(x a_i) = [C_k(x)]^k [a_1]^i$; that is, when we
add the "digit" a_i we multiply by k to "shift" and add i.

We also want to use the inverse, decoding function D_k from A_1^* to
A_k^*, but such a function D_k cannot be primitive recursive for the trivial
reason that it does not map all of A_k^* to A_k^*. This is easily overcome by
extending the domain of D_k and having $D_k(x) = D_k(a_1^{|x|})$ for all x in A_k^*.
The remaining and most important requirement is to have $D_k \circ C_k(x) = x$
for all x in A_k^*. To this end we first define primitive recursive functions
p, q, and r as follows: $p(\epsilon) = \epsilon$, $p(x a_i) = x a_i$ if $i \ne k$, and $p(x a_k) = p(x)$;
thus $p(x)$ is x with all consecutive a_k's removed from its end. We let
$q(\epsilon) = \epsilon$ and $q(x a_i) = q(x) a_1$; thus $q(x) = a_1^{|x|}$. And finally $r(\epsilon) = a_1$,
$r(x a_i) = x a_{i+1}$ if $i \ne k$, and $r(x a_k) = x a_k$; thus r is "almost" the
successor function $s(n) = n + 1$ on the natural numbers represented by
words in A_k^*. Now observe that D_k needs to be defined so that $D_k(\epsilon) = \epsilon$
and $D_k(x a_i)$ represents the integer "$D_k(x) + 1$." Thus $D_k(x a_i)$ is usually
$r(D_k(x))$, with the exceptions occurring when $D_k(x)$ ends in a string of
a_k's. In fact, if $D_k(x) = y a_j a_k^n$ with $j \ne k$ then we "count up one" with
$D_k(x a_i) = y a_{j+1} a_1^n$. Thus D_k is correctly defined by the following
primitive recursion:

$$D_k(\epsilon) = \epsilon$$

and

$$D_k(x a_i) = r(p(D_k(x))) q(D_k(x) - p(D_k(x))).$$

Therefore, we have our primitive recursive coding and decoding functions C_k and D_k which treat words in A_k^* as integers represented to
"base k." Since the underlying alphabet A_k is usually either clear from
the context or immaterial, we often drop the subscript and write C and D
instead of C_k and D_k.

Additional Exercises

*1.4.6 Let P be a predicate and define the predicate $\exists y \leq xP$ by $\exists y \leq xP(y, z_1, \ldots, z_n)$ if and only if there is some y with $|y| \leq |x|$ such that $P(y, z_1, \ldots, z_n)$. Show that if P is primitive recursive, then so is $\exists y \leq xP$.

*1.4.7

(a) Let P be a predicate and define the *function* $min_{y/x}P$ such that $min_{y/x}P(y, z_1, \ldots, z_n)$ is the shortest initial segment y of x such that $P(y, z_1, \ldots, z_n)$ if such a y exists, and xa_1 if no such y exists. Show that if P is primitive recursive then so is $min_{y/x}P$. (Why do we want the value xa_1 if no y exists?)

(b) Let P be a predicate and define the *function* $max_{y/x}P$ such that $max_{y/x}P(y, z_1, \ldots, z_n)$ is the longest initial segment y of x such that $P(y, z_1, \ldots, z_n)$ if such a y exists, and xa_1 if no such y exists. Show that if P is primitive recursive then so is $max_{y/x}P$.

We are interested in developing some primitive recursive functions over A_1^*; for the sake of convenience we identify A_1^* with N, the natural numbers. Thus we identify $a_1^{\ n}$ with n, and the base functions for the primitive recursive functions over "N" are simply the functions s, z, and $P_j^{\ n}$ where $s(n) = n + 1$, $z(n) = 0$, and $P_j^{\ n}(m_1, \ldots, m_n) = m_j$.

*1.4.8 Show that the following functions are primitive recursive over "N":

(a) $n + m$
(b) $n \cdot m$
(c) n^m.

*1.4.9 Show that the following predicates are primitive recursive over N:

(a) $n \leq m$, and $n < m$
(b) $n \mid m$ (n divides m evenly)
(c) $Prime(n)$ (n is a prime number).

1.4.10 Show that the following functions are primitive recursive over N:

(a) $pr(n) =$ the nth prime, with $pr(0) = 2$
(b) $(n)_m =$ the exponent of $pr(m)$ in the prime power expansion of n (for example, $(1960)_2 = (2^3 3^0 5^1 7^2)_2 = 1$).

*1.4.11 Suppose that for some $m > 0$, $k_0, \ldots, k_m$ and h are primitive recursive functions over N, and the function f is defined recursively as follows:

$$f(0, x_1, \ldots, x_n) = k_0(x_1, \ldots, x_n)$$
$$\vdots$$
$$f(m, x_1, \ldots, x_n) = k_m(x_1, \ldots, x_n)$$

and
$$f(y + 1, x_1, \ldots, x_n) = h(y, f(y, x_1, \ldots, x_n), x_1, \ldots, x_n)$$

for all $y \geq m$. Show that f is primitive recursive.

1.4.12 Let $max_{z \leq x}[g(z,y)]$ be the maximum of $\{g(0, y), \quad g(1, y), \ldots, g(x, y)\}$. Show that if g is primitive recursive then so is the function f defined by $f(x, y) = max_{z \leq x}[g(z, y)]$.

*1.4.13 Suppose that P is a predicate with characteristic function $c_P(y, x_1, \ldots, x_n)$. If f is a function P_f is the predicate whose characteristic function is $c_P(f(x_1, \ldots, x_n), x_1, \ldots, x_n)$. Show that if P and f are primitive recursive so is P_f. P_f is often written $P(f(x_1, \ldots, x_n), x_1, \ldots, x_n)$. If Q is a predicate, give an intuitive description of the predicate $\exists y \leq f(x_1, \ldots, x_n)Q$ and show that if Q is primitive recursive then so is $\exists y \leq f(x_1, \ldots, x_n)Q$.

1.4.14 Suppose that g and h are primitive recursive functions, and that $f(0, x) = g(x)$ and $f(y + 1, x) = f(y, h(x))$ for all x and y. Show that f is primitive recursive.

1.4.15 Let the function A be defined recursively by $A(0, n) = n+1$, $A(m+1, 0) = A(m, 1)$, and in general, $A(m+1, n+1) = A(m, A(m+1, n))$. Show that A is *not* primitive recursive by showing that for every primitive recursive function f there is some constant m such that for all $x_1, \ldots, x_n, f(x_1, \ldots, x_n) < A(m, max\{x_1, \ldots, x_n\})$. A is called the "Ackermann exponential."

1.5 THE PARTIAL RECURSIVE FUNCTIONS

Unfortunately, we have not yet accomplished our goal, stated at the beginning of Section 1.3, of characterizing the effectively computable functions, since not all of the effectively computable functions are primitive recursive. To see this, suppose we have a program or algorithm which itself lists programs or algorithms for effectively computable functions from $A_k{}^*$ to $A_k{}^*$. If all of the functions on this list are defined for all of their input arguments, then we can show how to give a new computable function from $A_k{}^*$ to $A_k{}^*$ which is not computed by any of the algorithms in the list. (The construction is a simple variation of that used by Cantor to show that no set can be put in one-to-one correspondence with its power set, or to show that there are uncountably many real numbers.) Let $f_0, f_1, f_2, \ldots$ be the functions computed by the algorithms in our list; then define the function g by

$$g(a_1{}^n) = f_n(a_1{}^n)a_1$$

and

$$g(x) = \epsilon$$

if $x \notin A_1{}^*$. Then g is effectively computable.

1.5.1 EXERCISE   Give an algorithm for computing g.

Also, g cannot appear in our list because if it were on the list, then for *some* m we would have $g = f_m$ and hence $g(a_1{}^m) = f_m(a_1{}^m)$; but by the definition of g we have $g(a_1{}^m) = f_m(a_1{}^m)a_1$.

It is not difficult to see how to generate systematically all possible ways of combining base functions via the operations of substitution and recursion, that is, to give a program which lists all of the primitive recursive programs. Therefore the construction above will apply, and so we see that there is an effectively computable function which is not primitive recursive. (Another way to see this fact is to work Exercise 1.4.15.) Moreover, *any* characterization of computable functions similar to the definition of the primitive recursive functions will be susceptible to the same attack. So we are doomed to leaving out some of the effectively computable functions if we attempt to characterize them by an effective list of programs which always return values for all possible inputs. However, we can see that the situation may not be hopeless. We are interested not only in computable functions, but in computable functions in conjunction with the algorithms which compute them, and anyone familiar with algorithms knows that not every algorithm produces an output (i.e., halts) on every possible input; sometimes algorithms end up in an infinite loop. This suggests that to include the functions computed by *all* algorithms it will be necessary to expand our horizons to include "partial functions."

To this end, we define a *partial function* from $(A_k{}^*)^n$ to $A_k{}^*$ to be a function whose domain is contained in $(A_k{}^*)^n$ and whose range is contained in $A_k{}^*$. We use $\phi, \psi, \ldots$ to denote partial functions and reserve $f, g, h, \ldots$ for functions known, intended, or asserted to be *total* functions, that is, functions with domain all of $(A_k{}^*)^n$. If ϕ and ψ are partial functions, then $\phi = \psi$ means that they are equal as sets of ordered pairs (i.e., they have the same domains and they have equal values at all arguments in their domain); $\phi(x) = \psi(x)$ means either that both sides are undefined or *divergent* (x is not in either function's domain) or that both sides are defined and that they are equal. Since we are thinking computationally, when a partial function is not defined on an argument we think of that as resulting from a nonterminating, "looping," or "divergent" computation. Note that asserting that a function is a partial function makes no statement as to whether or not it happens to be total.

Let $\psi, \theta_1, \ldots, \theta_m$ be given partial functions, then just as with total functions, we say that ϕ is obtained from them by *substitution* if

$$\phi(x_1, \ldots, x_n) = \psi(\theta_1(x_1, \ldots, x_n), \ldots, \theta_m(x_1, \ldots, x_n)),$$

and ϕ is obtained by *recursion* if

$$\phi(\epsilon, x_2, \ldots, x_n) = \psi(x_2, \ldots, x_n)$$

and

$$\phi(ya_i, x_2, \ldots, x_n) = \theta_i(y, \phi(y, x_2, \ldots, x_n), x_2, \ldots, x_n).$$

We have repeated these definitions in order to emphasize the interpretation that for the case of partial functions, the right-hand side of each equation above is defined *only* if all of the values within it are defined; thus for a definition by recursion, if $\phi(y, x_2, \ldots, x_n)$ is divergent, then $\phi(yz, x_2, \ldots, x_n)$ will also be divergent for all z.

Note that applying substitution and recursion to total functions always produces total functions. We now give a new operation on functions called minimization; this is a "search" operation which when applied to a total function can produce a partial function which is *not* total.

Scheme VI Let ψ be a given partial function; then ϕ is obtained from ψ by *minimization over $\{a_j\}^*$* if for all $x_1, \ldots, x_n$,

1. $\phi(x_1, \ldots, x_n)$ is defined if and only if there is an $m \in N$ such that for all p with $0 \le p \le m$, $\psi(a_j^p, x_1, \ldots, x_n)$ is defined and $\psi(a_j^m, x_1, \ldots, x_n) = \epsilon$.
2. When $\phi(x_1, \ldots, x_n)$ is defined (that is, when there is such an m), then $\phi(x_1, \ldots, x_n) = a_j^q$ where q is the least such m.

When ϕ is obtained from ψ by minimization over $\{a_j\}^*$ we write

$$\phi(x_1, \ldots, x_n) = min_j y[\psi(y, x_1, \ldots, x_n) = \epsilon],$$

and "$min_j y$" is read "the shortest y in $\{a_j\}^*$." If P is a predicate, $min_j y P(y, x_1, \ldots, x_n)$ means $min_j y[c_{notP}(y, x_1, \ldots, x_n) = \epsilon]$, which has the intuitively correct meaning of "the shortest y in $\{a_j\}^*$ such that $P(y, x_1, \ldots, x_n)$." Minimization allows us to perform "unbounded searches," and since such searches may fail to find what they are looking for, they do not always produce total functions, even when applied to total functions. Note that in Exercise 1.4.7 you showed how to perform bounded searches "primitive recursively," but it is not surprising that unbounded searches enable us to obtain functions which are not primitive recursive.

1.5.2 EXERCISE   Show how to use an algorithm for computing ψ to obtain one for computing $min_j y \psi$. (The reader familiar with high level programming languages such as ALGOL-60 should be able to give a solution to this exercise using a **while**-loop construction.)

We now define the class of partial recursive functions by using the operation of minimization to extend the class of primitive recursive functions. A function is a *partial recursive function* if it can be obtained from the base functions by (finitely many) applications of the operations of substitution, recursion, and minimization. A function is a *total recursive function* if it is a partial recursive function and it happens to be total. A predicate P is *recursive* if c_P is a total recursive function. In the following proposition we state some now obvious facts:

1.5.3 PROPOSITION   *Every primitive recursive function is a total recursive function, and the recursive predicates are closed under the Boolean operations of "and," "or," and "not," and also under the bounded quantifiers of Definition 1.4.3 and Exercise 1.4.6. Furthermore, the total recursive functions are closed under definitions by cases and the bounded searches of Exercise 1.4.7.*

From Exercises 1.3.3 and 1.5.2, it follows that all of the partial recursive functions are effectively computable. (In fact, given a definition of a partial recursive function we can easily produce an algorithm for computing it, so that just as we did with the primitive recursive functions we may regard the definition as providing a program for the function.) One of the main goals of this chapter is to provide evidence for the converse. Namely, that every effectively computable function is partial recursive. Our first step in this direction is to introduce an idealized programming language and computer and show that all partial recursive functions are computable on that machine. This "translation" of the partial recursive functions into a "machinelike" language shows how to compute the partial recursive functions using programs which may seem more familiar to computer scientists. Eventually, we prove that all partial functions computed by several different types of computational systems are partial recursive functions, providing important evidence for the Church-Turing Thesis which we discussed earlier.

1.5.4 EXERCISE Extend the definition of the primitive recursive programs given in Section 1.3 to include programs for all of the partial recursive functions.

1.6 RANDOM ACCESS MACHINE (RAM) PROGRAMS

A *random access machine* (RAM) is an idealized computer with random access memory; we fix an alphabet A_k and define the RAM for that alphabet. The RAM consists of a (potentially) infinite set of registers R1, R2, R3, . . . each of which can store any element of A_k^*. Any given program uses only the finite set of registers which it specifically names, and any given computation which halts uses only a finite set of words in A_k^*; thus any such computation needs only a finite amount of "hardware." (Note that our definition of the RAM, which follows, differs from other definitions used in the literature.) The RAM instructions use an infinite set of *line names* N0, N1, N2, . . . , and the RAM *instructions* are of the following seven types:

1_j.	X	**add**$_j$	Y	5.	X	**jmp**	X'
2.	X	**del**	Y	6_j.	X	Y **jmp**$_j$	X'
3.	X	**clr**	Y	7.	X	**continue**	
4.	X	$Y \leftarrow Z$					

where X is either a line name or nothing, Y and Z are register names, X' is a line name followed by an "a" or a "b" (e.g., N6a), and $1 \leq j \leq k$. Instructions of types 1 through 4 affect the contents of the registers in the following obvious ways: type 1_j adds a_j to the right end of the word in register Y; type 2 deletes the first (left end) letter of the word in Y (if any); type 3 changes the word in Y to ϵ; and type 4 copies the word in Z into Y leaving the word in Z unchanged. Instructions of types 5 and 6 are jumps which affect the order in which instructions are executed. Normally, instructions are executed sequentially in the order in which they are written. When a **jmp** Nia is executed, the next instruction to be executed is the first (closest) instruction above ("a" for above) bearing the line name Ni; **jmp** Nib goes to the first instruction below bearing line name Ni. Thus it is permissible and reasonable for several different instructions in a program to have the same line name, and Ni **jmp** Nia is not by itself an infinite loop. Type 6_j instructions are conditional jumps which are performed only if the first (left end) letter in the word in Y is a_j. Obviously, type 7 instructions are "no-ops," which do nothing.

A RAM *program* is a finite sequence of instructions such that each jump has a place to go (e.g., for Y **jmp**$_j$ N17b there must be a line with N17 below it), and such that the last instruction is a **continue**. A program *halts* if and when it reaches the final **continue** instruction. Thus we might have chosen to call the last instruction in a program a **halt** instruction, but for notational reasons it is convenient for us not to do so. For example, we have set things up so that we can use programs as subroutines without changing them at all.

A program **P** *computes* the partial function ϕ if when the initial contents of registers R1, R2, . . . , Rn are $x_1, x_2, . . . , x_n$ respectively, and the initial contents of all other registers (named in **P**) are ϵ, then (1) **P** halts eventually if and only if $\phi(x_1, . . . , x_n)$ is defined, and (2) if and when **P** halts, the final contents of R1 are $\phi(x_1, . . . , x_n)$. Thus for each RAM program **P** and for each $n \geq 1$, **P** computes the partial function $\phi_P{}^n$ of n arguments such that $\phi_P{}^n(x_1, . . . , x_n)$ is the final contents of R1 if and when **P** halts after being run on inputs $x_1, . . . , x_n$ initially in R1, . . . , Rn with the rest of the registers initially empty. A partial function is RAM-*computable* if some RAM program computes it.

The following are some very simple RAM programs:

1. **clr** R1	2. **add**$_j$ R1	3. R1 $\leftarrow$ Ri
continue	**continue**	**continue.**

The first one computes the erase function E; the second computes S_j; and the third computes $P_i{}^n$ for all $n \geq i \geq 1$. We have proved the following:

1.6.1 PROPOSITION *All of the base functions for the partial recursive functions are RAM-computable.*

Having shown that the base functions are RAM-computable, we proceed with propositions which show that in fact *all* of the partial recursive functions are RAM-computable.

1.6.2 PROPOSITION *If ψ, θ_1, . . . , θ_m are RAM-computable and ϕ is obtained from them by substitution, then ϕ is RAM-computable.*

Proof Let RAM programs **R**, **P**$_1$, . . . , **P**$_m$ compute ψ, θ_1, . . . , θ_m, respectively. Let n be the number of arguments in ϕ (and in the θ_i's), let q be the least integer greater than m and n and such that no register past Rq is named in **R**, **P**$_1$, . . . , **P**$_m$. Then the following RAM program computes ϕ:

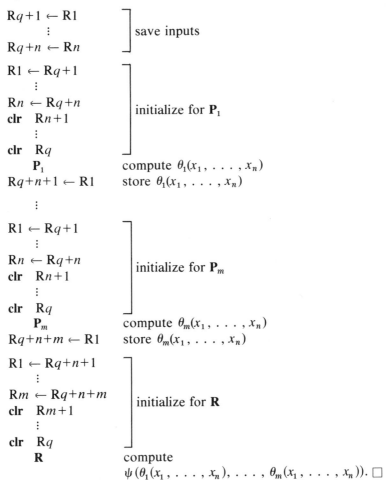

$$\left.\begin{array}{l} Rq+1 \leftarrow R1 \\ \quad\vdots \\ Rq+n \leftarrow Rn \end{array}\right\} \text{save inputs}$$

$$\left.\begin{array}{l} R1 \leftarrow Rq+1 \\ \quad\vdots \\ Rn \leftarrow Rq+n \\ \mathbf{clr} \quad Rn+1 \\ \quad\vdots \\ \mathbf{clr} \quad Rq \end{array}\right\} \text{initialize for } \mathbf{P}_1$$

$\mathbf{P}_1$ compute $\theta_1(x_1, \ldots, x_n)$
$Rq+n+1 \leftarrow R1$ store $\theta_1(x_1, \ldots, x_n)$

$$\vdots$$

$$\left.\begin{array}{l} R1 \leftarrow Rq+1 \\ \quad\vdots \\ Rn \leftarrow Rq+n \\ \mathbf{clr} \quad Rn+1 \\ \quad\vdots \\ \mathbf{clr} \quad Rq \end{array}\right\} \text{initialize for } \mathbf{P}_m$$

$\mathbf{P}_m$ compute $\theta_m(x_1, \ldots, x_n)$
$Rq+n+m \leftarrow R1$ store $\theta_m(x_1, \ldots, x_n)$

$$\left.\begin{array}{l} R1 \leftarrow Rq+n+1 \\ \quad\vdots \\ Rm \leftarrow Rq+n+m \\ \mathbf{clr} \quad Rm+1 \\ \quad\vdots \\ \mathbf{clr} \quad Rq \end{array}\right\} \text{initialize for } \mathbf{R}$$

$\mathbf{R}$ compute
$\psi(\theta_1(x_1, \ldots, x_n), \ldots, \theta_m(x_1, \ldots, x_n))$. $\square$

1.6.3 PROPOSITION   *If* $\psi, \theta_1, \ldots, \theta_k$ *are RAM-computable and* ϕ *is obtained from them by recursion, then* ϕ *is RAM-computable.*

1.6.4 EXERCISE   Prove the previous proposition.

1.6.5 PROPOSITION   *If* ψ *is RAM-computable and* ϕ *is obtained from it by minimization (over* $\{a_j\}^*$*), then* ϕ *is RAM-computable.*

Proof   We have $\phi(x_2, \ldots, x_n) = min_j y[\psi(y, x_2, \ldots, x_n) = \epsilon]$. Let $\mathbf{P}$ be a RAM program which computes ψ. Let n be the number of arguments in ψ and let q be the least integer greater than n such that no register past Rq is named in $\mathbf{P}$. Let Ni be a line name not used in $\mathbf{P}$, then the following program computes ϕ:

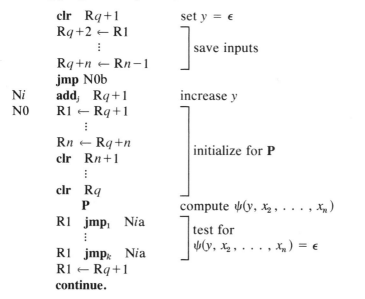

	clr $Rq+1$	set $y = \epsilon$
	$Rq+2 \leftarrow R1$	⎤
	⋮	save inputs
	$Rq+n \leftarrow Rn-1$	⎦
	jmp N0b	
Ni	**add**$_j$ $Rq+1$	increase y
N0	$R1 \leftarrow Rq+1$	⎤
	⋮	
	$Rn \leftarrow Rq+n$	initialize for **P**
	clr $Rn+1$	
	⋮	
	clr Rq	⎦
	P	compute $\psi(y, x_2, \ldots, x_n)$
	$R1$ **jmp**$_1$ Nia	⎤ test for
	⋮	$\psi(y, x_2, \ldots, x_n) = \epsilon$
	$R1$ **jmp**$_k$ Nia	⎦
	$R1 \leftarrow Rq+1$	
	continue.	□

The previous four propositions prove the following theorem:

1.6.6 THEOREM *Every partial recursive function is RAM-computable. Moreover, given a definition of a partial recursive function from the base functions via substitution, recursion, and minimization, we can effectively find a RAM program which computes the function.*

We close this section by showing that the RAM can be programmed with a smaller instruction set, a fact which will come in handy later.

1.6.7 PROPOSITION *Every RAM program can be effectively transformed into one which uses only instructions of types 1, 2, 6, and 7, and which is equivalent in the sense that it computes the same partial functions as the original program.*

Proof The strategy is to replace successively each of the other types of instructions by code using instructions not yet eliminated. Replace each instruction of the form X $Ri \leftarrow Ri$ by X **continue.** Let Rm be a register not named in the original program and let Nh, Ni, Nj_1, $\ldots$, Nj_k be line names not used in the original program. Then replace each instruction X $Rf \leftarrow Rg$ with f different from g by the following code, which "destructively" copies Rg into Rm and then "destructively" copies Rm back into Rf and Rg:

```
X        clr    Rf
         clr    Rm
         jmp    Nib
Nh       del    Rg
Ni       Rg     jmp₁    Nj₁b
                ⋮
         Rg     jmpₖ    Njₖb
         jmp    Nib
Nj₁      add₁   Rm
         jmp    Nha
                ⋮
Njₖ      addₖ   Rm
         jmp    Nha

Nh       del    Rm
Ni       Rm     jmp₁    Nj₁b
                ⋮
         Rm     jmpₖ    Njₖb
         jmp    Nib
Nj₁      add₁   Rf
         add₁   Rg
         jmp    Nha
                ⋮
Njₖ      addₖ   Rf
         addₖ   Rg
         jmp    Nha
Ni       continue.
```

copy Rg into Rm

copy Rm into Rf and Rg

The replacements for type 3 and 5 instructions are similar but simpler, and they are left as an exercise.

1.6.8 EXERCISE   Finish the proof of the previous proposition by showing how to replace type 3 and 5 instructions.   □

Note that if you finish the proof of Proposition 1.6.7 in a reasonable way, the program produced by the replacements will use only two more registers and $k + 2$ more line names than the original program, although it could have many more lines.

Additional Exercises

1.6.9   Calculate and justify a precise upper bound on the number of lines a program produced by the replacements for the proof of

Proposition 1.6.7 could have in terms of the number of lines in the original program.

1.6.10 Show that type 1, 2, 6, and 7 instructions in fact form a minimal set of instructions for the RAM by showing that if we eliminate any one of these types of instructions we will no longer have programs for computing all of the RAM-computable functions.

1.7 TURING MACHINES

A *Turing machine* is an idealized computer with associative memory; it contains a two-way (potentially) infinite tape divided into squares and a finite control device with a read-write head which moves along the tape. The Turing machine uses some finite *alphabet* A_k of tape symbols; a_k is usually taken to be a blank, and is often denoted by B. The finite control device has a finite set of *internal states*, $\{0, 1, \ldots, p\}$. The *instructions* in the control device are quintuples of the form $i\, a_j\, a_n\, D\, m$ where $0 \leq i, m \leq p$ and $1 \leq j, n \leq k$, and D stands for either R or L; the interpretation of an instruction $i\, a_j\, a_n\, D\, m$ is "if in state i looking at symbol a_j on the tape, replace it by a_n, move one square in direction D along the tape, and go into state m." Turing machines are generally assumed to be *deterministic*; that is, for any given state i and symbol a_j there is at most one instruction beginning $i\, a_j\, \ldots$. (We have occasion to consider nondeterministic Turing machines in Chapter 7 of this book, however.) A Turing machine *halts* when it is in a state i looking at a symbol a_j and there is no instruction beginning $i\, a_j\, \ldots$.

Let ϕ be a partial function over $A_k{}^*$. A Turing machine **T** *computes* ϕ if the alphabet of **T** is A_{k+2} with a_{k+1} a comma (,) and a_{k+2} a blank (B), and if *when* **T** is given a tape containing "$x_1, \ldots, x_n$" (the rest blank; note that the ,'s are part of this input string given to **T**) and started in state 0 over the left end of the input string, *then* (1) **T** halts eventually if and only if $\phi(x_1, \ldots, x_n)$ is defined, and (2) if and when **T** halts, the final content of the tape is $\phi(x_1, \ldots, x_n)$ with the remaining tape blank.

Equivalently, for each integer $n \geq 1$, each Turing machine **T** computes *some* partial function $\phi_T{}^n$ of n arguments defined as follows: *if* it happens that **T** halts on an input of the form $x_1, x_2, \ldots, x_n$ with each x_i in $A_k{}^*$ and halts with the nonblank portion of the tape a consecutive segment of tape containing a string y in $A_k{}^*$, *then* $x_1, \ldots, x_n$ is in the domain of $\phi_T{}^n$ and $\phi_T{}^n(x_1, \ldots, x_n) = y$. A partial function is *TM-computable* if some Turing machine computes it.

To make it easier to program some Turing machines, we introduce *state transition diagrams* for Turing machines: an example is given in

Figure 1.7.1. The diagrams consist of labeled circles connected by labeled arrows; each state of the machine corresponds to one of the circles, and the remaining circles represent halting conditions; arrows correspond to transitions (instructions). An instruction $i\,a\,b\,D\,j$ is represented by an arrow labeled a/b from the circle for state i to that for state j; in addition, D is placed inside the circle for state j. If there is no instruction beginning $i\,a$ . . . then this is represented by an arrow labelled a from the circle for state i to one of the circles for halting conditions, which has an H inside it. The initial state, 0, is indicated by an arrow from the word "start" to the circle which represents it. The way these diagrams are "read" is that an arrow labeled a/b from one state circle to another with D inside of it means that when the machine is in the first state looking at an a it changes it to a b and moves in direction D on entering the state at the head of the arrow. Of course this interpretation makes sense only when each state circle has in it at most one of the direction symbols R and L.

We simplify these diagrams by adopting the convention that an arrow labeled with only one letter a is interpreted as if it were labeled with a/a (i.e., leave the a alone) and that a state circle with no arrow labeled by a/x for any symbol x coming out of it is interpreted as if there were an arrow from that circle back to itself labeled a/a (i.e., skip over the a moving in the direction in the state, and stay in the same state). Figure 1.7.1 gives a state transition diagram for the Turing machine with alphabet $\{a, b, B\}$, states $\{0, 1, 2\}$, and instructions

$$
\begin{array}{lll}
0\,a\,a\,R\,0 & 1\,a\,a\,R\,1 & 2\,a\,a\,L\,2 \\
0\,b\,b\,R\,1 & 1\,b\,a\,L\,2 & 2\,b\,b\,L\,2 \\
 & & 2\,B\,a\,R\,0.
\end{array}
$$

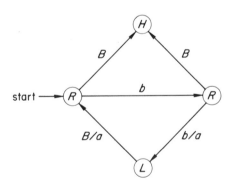

FIGURE 1.7.1   A state transition diagram of a simple Turing machine.

Note that such a machine moves in the direction indicated within a state each time it enters that state *except* the first time it enters the initial state; that is, we think of the machine as being in the initial state already when it begins its computation.

1.7.2 EXERCISE Describe what the Turing machine in Figure 1.7.1 does when it is started in its initial state over the left end of an input string in $\{a, b\}^*$.

This system of state transition diagrams often allows us to draw fairly simple pictures of Turing machines which are much easier to read than a list of instructions would be. However, the system does have a drawback; because of our convention of putting the direction of the move inside the next state, not every Turing machine has a state transition diagram. We have allowed this because, in fact, many of the Turing machines which people actually produce do have state transition diagrams of this simple type. And in any case, every Turing machine can easily be converted into an equivalent one which does have a state transition diagram by splitting each state into two, one that always moves right when entered and one that always moves left when entered. Since we do not need this result, we do not even give a formal statement of it, but we do leave it as an optional exercise (Exercise 1.7.9). You will get practice in state transition diagrams from the proof of the next theorem, which shows how to translate RAM programs into Turing machines.

1.7.3 THEOREM *Every RAM-computable function is TM-computable. Moreover, given a RAM program we can effectively find a Turing machine which computes the same partial functions.*

Proof Let **P** be a RAM program which uses no registers other than R1, . . . , Rm. If the contents of these registers are $r1, \ldots, rm$ respectively, then we represent these contents on the Turing machine tape by the string $r1, \ldots, rm$, (the rest blank, the ,'s are part of the string), and we show how to draw a section of a Turing machine diagram which will simulate the action of any given RAM instruction on this representation. By Proposition 1.6.7, we may assume that the RAM program **P** contains only instructions of types 1, 2, 6, and 7. If **P** has n instructions then our Turing machine is built of n blocks connected for the same "flow of control" as the n instructions in **P**; the jth block in the Turing machine diagram simulates the action of the jth instruction in **P** on our tape representation of the RAM registers. What is in the jth block of the Turing machine diagram thus depends on the type of the jth instruction in **P** as well as on the register (if any) to which the instruction refers. For

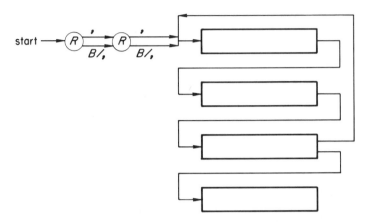

FIGURE 1.7.4   The block structure of a Turing machine to simulate a simple RAM program.

example, Figure 1.7.4 gives the block structure of the Turing machine corresponding to the RAM program below, together with some states to "initialize" the input tape:

$$N0 \quad \textbf{add}_1 \quad R1$$
$$\textbf{del} \quad R2$$
$$R2 \quad \textbf{jmp}_1 \quad N0a$$
$$\textbf{continue.}$$

In the machine for our general program **P**, the chain of "initialization" states leading into the first block contains m states connected with the same transitions as in Figure 1.7.4, which makes sure that there are at least m ,'s setting off the contents of R1, . . . , Rm on the tape even if the initial contents of some of these are empty. The reason for this is that if the machine is computing a function of t arguments with $t < m$, input strings are of the form $x_1, \ldots, x_t$ and the initial value of $rt + 1$, . . . , rm is ϵ, and so we need $m+1-t$,'s at the end of the input string to set these off. What goes into each of the blocks for our general program is given in Figure 1.7.5. To simplify our *pictures*, we assume that the RAM alphabet is $A_2 = \{a_1 = 0, a_2 = 1\}$; then our Turing machine alphabet is $\{0, 1, , , B\}$. The modifications for larger alphabets are obvious. If the jth instruction in **P** is $\textbf{add}_i$ Rq then the jth block of our Turing machine is given in Figure 1.7.5a. If the jth instruction in **P** is **del** Rq then the jth block is given in Figure 1.7.5b. If the jth instruction is Rq $\textbf{jmp}_i Z$ then the jth block is given in Figure 1.7.5c. In each of these cases the block first finds the representation of the contents of Rq on the tape

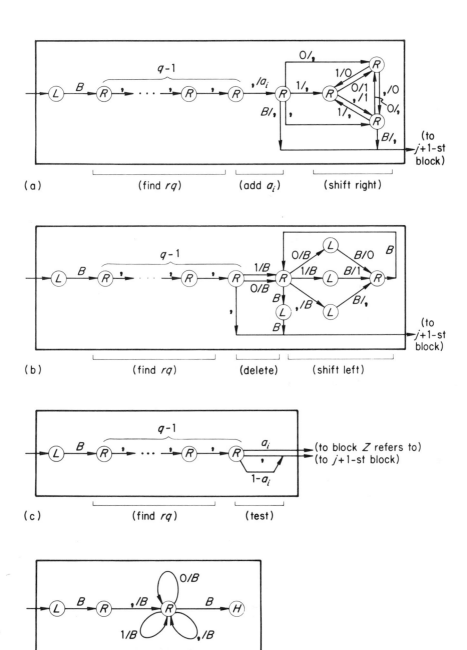

FIGURE 1.7.5 Turing machine blocks for simulating various RAM instructions.

and then performs the required operation. The nth block, for the final **continue** in **P**, is given in Figure 1.7.5d; it erases all but the contents of $R1$ from the tape before halting.

1.7.6 EXERCISE   Suppose the jth instruction in **P** is **continue** and $j <$ n. In less than ten seconds, give the jth block of our Turing machine.

This completes the proof of the theorem.   □

Notice that the Turing machine produced by the translation in Theorem 1.7.5 has the following nice property: the Turing machine never moves left of the blank square immediately to the left of its starting position. Thus Turing machine tape need only be unbounded to the right. When our circle of translations is completed in the next section, this will yield that every Turing machine computable function is computable by a Turing machine which never moves more than one square to the left of its starting position.

1.7.7 EXERCISE   Show how to convert every Turing machine into one which never moves more than one square to the left of its starting position and which, for each n, computes the same partial function of n arguments as the original Turing machine.   *Hint*: your converted Turing machine will probably add special markers on either end of the portion of the tape visited.

Additional Exercises

1.7.8   Suppose the program **P** in the proof of Theorem 1.7.3 contained instructions of types 3 and 5. Show what to put in the corresponding block of our Turing machine to simulate these instructions.

1.7.9   Show how to convert any Turing machine into an equivalent one which has a state transition diagram, and prove that your construction works. Note that as part of your proof you need to give a precise definition of what you mean when you say that the two Turing machines are "equivalent."

1.8   MARKOV ALGORITHMS

Markov algorithms are formal symbol manipulation schemes which might be considered deterministic versions of formal grammars. Before we give the formal definition of Markov algorithms, we illustrate how they work with a very simple example. Our example Markov algorithm consists of the following *sequence* of productions, or rewrite rules:

$$ac \rightarrow ca$$
$$bc \rightarrow cb$$
$$c \rightarrow a$$
$$bb \rightarrow bc.$$

Suppose we are given a word in $\{a, b\}^*$, and we wish to see how this algorithm operates on that word. We take the first production whose left-hand side occurs in the word and apply that production as far to the left as possible. Thus for a word in $\{a, b\}^*$ our algorithm changes the leftmost occurrence of bb to bc. Continuing in the same manner, the algorithm will then move the c to the left, one letter at a time, until it is at the left end of the word; then it will change the c to an a, after which the whole process will start all over again. Therefore given m and n in N, our algorithm transforms $a^m b^{n+1}$ into $a^{m+n}b$, at which point none of the productions apply and the algorithm halts.

1.8.1 EXERCISE Describe how the example Markov algorithm above has transformed any word in $\{a, b\}^*$ when it eventually halts.

We now give the formal definition. A *Markov algorithm* (MA) on the alphabet A_k is a finite ordered sequence of *productions*

$$p_1 \rightarrow_{(t)} q_1$$
$$\vdots$$
$$p_n \rightarrow_{(t)} q_n$$

where each p_i and q_i is a word in A_k^* and (t) means that each production may or may not have a t marking the arrow; $p \rightarrow q$ is called a *simple* production, and $p \rightarrow_t q$ is called a *terminating* production.

If **M** is a Markov algorithm on A_k and x and y are in A_k^*, we say that **M** *transforms x directly into y*, written **M**:$x \rightarrow y$, if p_i is the first of $p_1, \ldots, p_n$ occurring in x and y is the result of replacing the leftmost occurrence of p_i in x by q_i, where $p_i \rightarrow q_i$ is a simple production of **M**. **M** *transforms x directly into y terminally*, written **M**:$x \rightarrow_t y$, if p_i is the first of $p_1, \ldots, p_n$ occurring in x and y is the result of replacing the leftmost occurrence of p_i in x by q_i, where $p_i \rightarrow_t q_i$ is a terminating production of **M**. Furthermore, **M** *transforms x into y in m steps*, written **M**:$x \underset{m}{\Rightarrow} y$, is defined by induction on m:

$$\mathbf{M}:x \underset{0}{\Rightarrow} y \quad \text{iff} \quad x = y$$

and

$$\mathbf{M}:x \underset{m+1}{\Rightarrow} y \quad \text{iff} \quad \text{for some } z \in A_k^*, \quad \mathbf{M}:x \underset{m}{\Rightarrow} z \quad \text{and} \quad \mathbf{M}:z \rightarrow y.$$

M *transforms* x *into* y *in* $m+1$ *steps terminally,* written $\mathbf{M}:x \underset{m+1}{\Rightarrow}_t y$, if and only if for some $z \in A_k{}^*, \mathbf{M}:x \underset{m}{\Rightarrow} z$ and $\mathbf{M}:z \to_t y$. $\mathbf{M}(x) = y$ if and only if either $\mathbf{M}:x \underset{m}{\Rightarrow}_t y$ for some $m \in N$, or $\mathbf{M}:x \underset{m}{\Rightarrow} y$ for some $m \in N$ and no p_i occurs in y. Finally, we say that $\mathbf{M}(x)$ is *defined* if and only if $\mathbf{M}(x) = y$ for some y.

We now define how Markov algorithms are used to compute partial functions. Let ϕ be a partial function of m arguments over $A_k{}^*$, and let $n > k$ be such that , is in $A_n - A_k$. A Markov algorithm **M** on A_n *computes* ϕ if for all $x_1, \ldots, x_m$ in $A_k{}^*$,

$$\phi(x_1, \ldots, x_m) = \mathbf{M}(x_1, \ldots, x_m)$$

where $x_1, \ldots, x_m$ stands for the single string including the commas. That is, **M** computes ϕ if (1) $\mathbf{M}(x_1, \ldots, x_m)$ is defined if and only if $\phi(x_1, \ldots, x_m)$ is defined, and (2) if $\mathbf{M}(x_1, \ldots, x_m)$ is defined then it equals $\phi(x_1, \ldots, x_m)$. A partial function is MA-*computable* if some Markov algorithm computes it.

The Markov algorithms on $\{a_1, a_2, , , \$\}$ below compute S_2 and $P_2{}^2$ over $A_2{}^*$, respectively; note the importance of the ordering of the productions.

1.	$\$a_1 \to a_1\$$	2. $\$a_1 \to \$$
	$\$a_2 \to a_2\$$	$\$a_2 \to \$$
	$\$ \to_t a_2$	$\$, \to_t \epsilon$
	$\epsilon \to \$$	$\epsilon \to \$$

Use these examples to make sure you understand the definitions above.

1.8.2 EXERCISE   If one of the Markov algorithms above is applied to $x \in A_2{}^*$, which production is applied *first*, and why?

1.8.3 THEOREM   *Every TM-computable function is MA-computable. Moreover, given a Turing machine we can effectively find a Markov algorithm which computes the same partial functions.*

Proof   What we prove is actually even stronger than the statement of the theorem. Given a Turing machine, we produce a Markov algorithm which not only computes the same partial functions but actually simulates the Turing machine in a step-by-step fashion in such a way that on any input x in the alphabet of the Turing machine, the Markov algorithm halts if and only if the Turing machine does, and furthermore when the Markov algorithm does halt, the resulting string is the nonblank contents of the Turing machine tape when the Turing machine halts after being given the same input. Let T be a Turing machine with alphabet A_k and states $\{0, \ldots, p\}$. Our Markov algorithm M to simulate T will use the alphabet

$$A_{k+p+5} = \{a_1, \ldots, a_k, r_0, \ldots, r_p, r_{p+1}, r_{p+2}, \#, \$\}.$$

Suppose that the Turing machine is in state i reading symbol a_j and that xa_jy is the contents of a section of the Turing machine tape which includes all nonblank squares; then our simulation of T represents this situation of T being in state i and looking at the a_j on this tape by the string $\#xr_ia_jy\#$. Thus we have inserted a symbol indicating what state T is in directly to the left of the tape symbol T is currently scanning; such a representation is called an *instantaneous description* of T at this point in its computation (a notion used several times later in this book). Notice that we have put $\#$'s on either end of the instantaneous description to serve as convenient end markers. Our Markov algorithm M first takes an input word and converts it into an instantaneous description of the initial configuration of T on that input, then M simulates T by successively transforming each instantaneous description of T directly into the next instantaneous description, and if and when T halts, M performs some "clean-up" and halts itself.

We begin the sequence of productions in M with productions

$$r_ia_ja_1 \rightarrow a_nr_ma_1$$
$$\vdots$$
$$r_ia_ja_k \rightarrow a_nr_ma_k$$
$$r_ia_j\# \rightarrow a_nr_mB\#$$

for each instruction $i\ a_j\ a_n\ R\ m$ in T; recall that B is a blank. For each instruction $i\ a_j\ a_n\ L\ m$ in T we have productions

$$a_1r_ia_j \rightarrow r_ma_1a_n$$
$$\vdots$$
$$a_kr_ia_j \rightarrow r_ma_ka_n$$
$$\#r_ia_j \rightarrow \#r_mBa_n.$$

The productions so far have M transform each instantaneous description directly into the next. To handle the halting of T, we add the following productions: for each pair $i\ a_j$ such that T halts when in state i looking at a_j we add the production

$$r_ia_j \rightarrow r_{p+1}a_j,$$

then we add the productions

$$a_1 r_{p+1} \rightarrow r_{p+1} a_1$$
$$\vdots$$
$$a_k r_{p+1} \rightarrow r_{p+1} a_k$$
$$\# r_{p+1} \rightarrow r_{p+2}$$
$$r_{p+2} a_1 \rightarrow a_1 r_{p+2}$$
$$\vdots$$
$$r_{p+2} a_{k-1} \rightarrow a_{k-1} r_{p+2}$$
$$r_{p+2} B \rightarrow r_{p+2}$$
$$r_{p+2} \# \rightarrow_t \epsilon;$$

(recall that $a_k = B$).

1.8.4 EXERCISE   Describe exactly how the last group of productions "cleans up" the output tape of **T**.

It is important to notice (for later use in Sections 5 and 6 of Chapter 2), that for all of the productions we have given so far, their *order* is completely unimportant. Given *any* instantaneous description of **T**, *exactly one* of these productions will apply in exactly one location, and it will produce either the next instantaneous description of **T**, or if **T** has halted, it will begin the "clean-up" phase.

Finally, to complete the proof of the theorem we give some "initialization" productions which take any input string in $A_k{}^*$ and transform it into an instantaneous description of the initial configuration of **T**. These productions are

$$\epsilon \rightarrow \# r_0 \$$$

and the group

$$\$ a_1 \rightarrow a_1 \$$$
$$\vdots$$
$$\$ a_k \rightarrow a_k \$$$
$$\$ \rightarrow \#.$$

Here the order of the productions becomes important; the first one must be the *last* production in **M** so that it is applied only once (at the beginning of **M**'s "computation"), and the last one in the group above must be after the others in that group; also, to see that **M** actually works as claimed, it is simplest to have the group of productions above at the beginning of **M**. You should now be able to easily verify that **M** works as we have claimed.   □

So far we have proved that every partial recursive function is RAM-computable, that every RAM-computable function is TM-computable, and that every TM-computable function is MA-computable. To show that all of these apparently different definitions of classes of computable functions really do give the same class of functions, we must perform one more reduction. This reduction is the one which shows that every MA-computable function is a partial recursive function, and it is perhaps the most difficult one we do. For the purpose of proving this reduction, let M be a fixed Markov algorithm on the alphabet A_k. We begin by showing that certain functions and predicates related to the action of M are primitive recursive.

1.8.5 PROPOSITION *The following are primitive recursive:*

1. $Oc(x, y)$ where $Oc(x, y)$ iff x occurs in y.
2. $u(x, z) = $ the word to the left of the leftmost occurrence of x in z (if $Oc(x, z)$, otherwise *anything*).
3. $v(x, z) = $ the word to the right of the leftmost occurrence of x in z (if $Oc(x, z)$, otherwise *anything*).
4. $rep(x, y, z) = $ the result of replacing the leftmost occurrence of x by y in z (if $Oc(x, z)$, otherwise *anything*).

Proof

1. $Oc(x, y)$ iff $\exists z/y \; \exists w/y \; [z = wx]$.
2. $u(x, z) = min \; y/z \; \exists w/z \; [yx = w]$.
3. $v(x, z) = z - u(x, z)x$.
4. $rep(x, y, z) = u(x, z)yv(x, z)$. □

1.8.6 PROPOSITION *For any Markov algorithm M, the following are primitive recursive:*

1. M such that $M(x, y) = z$ if $M : x \Rightarrow_{(t)|y|} z$ (otherwise we do not care what M yields);
2. $Term_M(x)$ iff M transforms x directly into y terminally for some y;
3. $Stop_M(x, y)$ iff M halts on x in $|y|$ steps.

Proof

1. $M(x, \epsilon) = x$ and

$$
M(x, ya_i) = \begin{cases}
rep(p_1, q_1, M(x, y)) & \text{if} \quad Oc(p_1, M(x, y)) \\
rep(p_2, q_2, M(x, y)) & \text{if} \quad Oc(p_2, M(x, y)) \; and \\
& \qquad not \; Oc(p_1, M(x, y)) \\
& \vdots \\
rep(p_n, q_n, M(x, y)) & \text{if} \quad Oc(p_n, M(x, y)) \; and \\
& \qquad not \; Oc(p_1, M(x, y)) \; \ldots \; and \\
& \qquad not \; Oc(p_{n-1}, M(x, y)) \\
M(x, y) & \text{otherwise;}
\end{cases}
$$

2. Let $p_{i_1} \to_t q_{i_1}, \ldots, p_{i_m} \to_t q_{i_m}$ be the terminating productions of **M**. Then $Term_M(x)$ iff

$$[not\ Oc(p_1, x)\ \ and \ldots and\ \ not\ Oc(p_{i_1-1}, x)\ \ and\ \ Oc(p_{i_1}, x)]$$
$$or$$
$$\vdots$$
$$or$$
$$[not\ Oc(p_1, x)\ \ and \ldots and\ \ not\ Oc(p_{i_m-1}, x)\ \ and\ \ Oc(p_{i_m}, x)];$$

3. $Stop_M(x, \epsilon)$ iff $not\ Oc(p_1, x)$ $and \ldots and$ $not\ Oc(p_n, x)$, and

$$Stop_M(x, ya_i)\ \ iff\ \ Term_M(M(x, y))\ \ or$$
$$[not\ Oc(p_1, M(x, ya_i))\ \ and \ldots and\ \ not\ Oc(p_n, M(x, ya_i))].$$
$$\square$$

Now note that for all x, $\mathbf{M}(x) = M(x, min_1 y[Stop_M(x, y)])$. To complete our final "translation" we need to be able to give **M** as input the single string $x_1, \ldots, x_m$ which represents the m arguments $x_1, \ldots, x_m$. To this end we define the helpful primitive recursive function I_m of m arguments such that

$$I_m(x_1, \ldots, x_m) = x_1, \ldots, x_m = con_{2m-1}(x_1, ,, \ldots, ,, x_m).$$

1.8.7 THEOREM   *Every MA-computable function is a partial recursive function. Moreover, given a Markov algorithm and any $m \in N$ we can effectively find a definition (from the base functions via substitution, recursion, and minimization) of the partial function of m arguments which the Markov algorithm computes.*

Proof   Let ϕ be a partial function of m arguments computed by the Markov algorithm **M**. Then for all $x_1, \ldots, x_m$

$$\phi(x_1, \ldots, x_m) = M(I_m(x_1, \ldots, x_m), min_1 y[Stop_M(I_m(x_1, \ldots, x_m), y)]). \quad \square$$

It is important to notice that this proof also shows that every partial recursive function ϕ can be obtained in the form

$$\phi = f \circ min_1 g$$

where f and g are primitive recursive. Not only is this fact of aesthetic interest, but we make use of it many times later in this book.

To summarize the translation results of Theorems 1.6.6, 1.7.3, 1.8.3, and 1.8.7, we have the following theorem:

1.8.8 THEOREM   *The following classes of partial functions are all the same:*

1. Partial recursive functions

2. *Random access machine computable functions*
3. *Turing machine computable functions*
4. *Markov algorithm computable functions*

Moreover, these equivalences are effective in that given a program for a partial function in any one of these four formalisms we can effectively translate it into a program for the same function in any of the other formalisms.

The preceding theorem shows that various attempts to give a precise definition of the intuitively understood class of effectively computable functions actually all define the same class of functions. The theorem is only a part of the evidence for the Church-Turing Thesis. In addition, every function believed on intuitive grounds by computer scientists and mathematicians to be effectively computable is in fact a partial recursive function. You should not find this surprising if you realize that all of the many standard programming languages which have been devised can be translated into RAM programs.

The observant reader has no doubt noticed that if we follow one definition of a function around the circle of translations we have given, the alphabet serving as a base for the function enlarges. This bothersome situation can be eliminated (for all but the one-letter alphabet A_1) by extensive use of coding and decoding to simulate having a larger alphabet when necessary. We have already seen some use of such coding techniques, and the next chapter contains many more. Any reader actually interested in performing this task is, somewhat reluctantly, encouraged to do so.

Additional Exercises

The *single register machine* (*SRM*) on A_k consists of a single register which can store any word in A_k^*. Its instructions use *line names* N0, N1, N2, ... The *instructions* are of the following three types:

1_j.	X	**add**$_j$	(add a_j to the right end of the register)
2.	X	**del**	(delete the left end symbol in the register, if any)
3.	X	**jmp**$_j$ X'	(jump to X' if the register begins with a_j)

where X is a line name and X' is a line name followed by an "a" (for above) or a "b" (for below). An SRM *program* is a finite sequence of instructions such that each jump has a place to jump to; programs *halt* when they exit from their last instruction looking for a next instruction.

If ϕ is a partial function on $A_k{}^*$, then an SRM program $\mathbf{P}$ on $A_{k+1} = A_k \cup \{,\}$ computes ϕ if when the initial contents of the register are "$x_1, \ldots, x_n$" with the x_i's in $A_k{}^*$, then (1) $\mathbf{P}$ halts if and only if $\phi(x_1, \ldots, x_n)$ is defined, and (2) if and when $\mathbf{P}$ halts, the final contents of the register are $\phi(x_1, \ldots, x_n)$.

1.8.9 Write a transfer routine for the SRM. If the contents of the register are $x_1, x_2, \ldots, x_n$ with the x_i's in $A_k{}^*$, then this routine should leave $x_2, \ldots, x_n, x_1$ in the register.

1.8.10 Show that every RAM-computable function is SRM-computable.

1.8.11 Show that every SRM-computable function is TM-computable.

1.9 THE COMPLEXITY OF TRANSLATED PROGRAMS

As part of our evidence for the Church-Turing Thesis, in Sections 1.5–1.8 we gave four "programming systems" for the effectively computable functions, and we proved that these programming systems are all equivalent in the sense that they all compute the same class of functions, the partial recursive functions. Moreover, these equivalences were established by giving simple (although perhaps tedious) effective "translations" of programs in one system to equivalent programs in another system. In this section we examine the efficiency, or *computational complexity* of the programs produced by the translations from Sections 1.6–1.8. What we mean here by the computational complexity of a program is not the complexity of the program per se (that is, the complexity of its "structure"), but rather the program's *use of computational resources* such as time or memory space for which one might be "charged" when running the program. Since a program's use of such resources depends on its actual execution (that is, on its computation) on a *particular input*, the complexity of a program in this sense must be a function of the inputs to the program. The goal of this section is to show that the translations from Sections 1.6–1.8 produce translated programs whose complexities, in terms of the general theory of algorithms, are very little more than the complexities of the programs which were translated. These results are used in Chapter 6, and are important to an understanding of Chapters 5 and 7.

Our first task is to define some simple "complexity measures" for the programming systems from Sections 1.5–1.8; we shall use a couple of convenient abbreviations for this purpose: if $x_1, \ldots, x_n$ are strings in some $A_k{}^*$ then $\bar{x}$ will stand for the n-tuple $x_1, \ldots, x_n$ and $|\bar{x}|$ will stand for the sum of the lengths of x_1 through x_n, $\Sigma_{1 \le i \le n} |x_i|$. Let $\mathbf{P}$ be a RAM program over alphabet A_k; we define the partial function

$$\text{RAM}space_P:(A_k{}^*)^n \to N(=A_1{}^*)$$

as follows: $\text{RAM}space_P(\vec{x})$ is the maximum total length of the contents of the RAM registers at any point in **P**'s computation on inputs $\vec{x}$ if that computation halts, and $\text{RAM}space_P(\vec{x})$ is undefined if **P**'s computation on inputs $\vec{x}$ does not halt. First, it is not obvious that $\text{RAM}space_P$ is a partial recursive function since a program may well get into an "infinite loop" which does not generate arbitrarily long strings in the registers; but such infinite loops can actually be detected by monitoring the program's execution. Specifically, notice that if **P** is a RAM program over A_k with m instructions (and hence using at most m registers) then there are less than $(k+1)^{mn}$ different possible contents of the RAM registers used by **P** with total length no greater than n; therefore if **P** runs for more than $m(k+1)^{mn}$ steps (instruction executions) such that the total length of the register contents does not exceed n then **P** *must* be in an infinite loop.

1.9.1 EXERCISE Using the comments from the previous sentence, show that $\text{RAM}space_P$ is in fact a partial recursive function.

Let **P** be a RAM program and define $\text{RAM}time_P(\vec{x})$ to be $|\vec{x}|$ plus the number of instruction executions in **P**'s computation on inputs $\vec{x}$ if that computation halts, and undefined if the computation does not halt. When dealing with computational complexity it is particularly convenient to use ∞ as the "value" of a partial function which is undefined; thus we could simply define $\text{RAM}time_P(\vec{x})$ as $|\vec{x}|$ plus the number of instruction executions in **P**'s computation on inputs $\vec{x}$. If you are wondering why we included $|\vec{x}|$ in our definition of $\text{RAM}time$, recall that we never specified just how the inputs $\vec{x}$ got into the RAM registers; since they must get there somehow, it seems only reasonable that $\text{RAM}time$ "charge" $|\vec{x}|$ for the performance of this service.

1.9.2 EXERCISE Show that $\text{RAM}time_P$ is a partial recursive function.

There are some obvious relationships between $\text{RAM}space_P$ and $\text{RAM}time_P$. Since each RAM instruction can add at most one character to the total contents of the RAM registers, obviously $\text{RAM}space_P(\vec{x}) \leq \text{RAM}time_P(\vec{x})$ for all $\vec{x}$. By the observations in the sentence preceding Exercise 1.9.1, there is a positive integer constant k_P (which depends on the program **P**) such that $\text{RAM}time_P(\vec{x}) \leq (k_P)^{\text{RAM}space_P(\vec{x})+1}$ for all $\vec{x}$.

1.9.3 EXERCISE Verify the previous assertion.

Let **T** be any Turing machine. Define $\text{TM}space_T(\vec{x})$ to be the total number of tape squares used by **T** (that is, actually visited by **T** *or* originally occupied by $\vec{x}$) in its computation on inputs $\vec{x}$ if that computation halts, and undefined if the computation does not halt. Define $\text{TM}time_T(\vec{x})$ to be $|\vec{x}|$ plus the number of instruction executions in T's computation on inputs $\vec{x}$. $\text{TM}space$ and $\text{TM}time$ are analogous to the space and time measures for RAM programs defined above.

1.9.4 EXERCISE

(a) Show that $\text{TM}space_T$ and $\text{TM}time_T$ are both partial recursive functions;

(b) show that $\text{TM}space_T(\vec{x}) \leq \text{TM}time_T(\vec{x})$ for all $\vec{x}$; and

(c) show that there is a positive integer constant k_T (which depends on T) such that $\text{TM}time_T(\vec{x}) \leq (k_T)^{\text{TM}space_T(\vec{x})}$ for all $\vec{x}$.

Let **M** be any Markov algorithm. Define $\text{MA}space_M(\vec{x})$ to be the maximum length of a string in **M**'s computation on inputs $\vec{x}$ if that computation halts, and undefined if the computation does not halt; in terms of Proposition 1.8.6, if there is a z such that $Stop_M(\vec{x}, z)$ then

$$\text{MA}space_M(\vec{x}) = max\{|M(\vec{x}, y)| : |y| \leq min_1 z[Stop_M(\vec{x}, z)]\}.$$

Define $\text{MA}time_M(\vec{x})$ to be $|\vec{x}|$ plus the number of steps in **M**'s computation on inputs $\vec{x}$; that is, $\text{MA}time_M(\vec{x}) = |\vec{x}| + |min_1 z[Stop_M(\vec{x}, z)]|$.

1.9.5 EXERCISE

(a) Show that $\text{MA}space_M$ and $\text{MA}time_M$ are both partial recursive functions.

(b) Show that there is a positive integer constant k_M (which depends on **M**) such that $\text{MA}space_M(\vec{x}) \leq k_M \text{MA}time_M(\vec{x})$ for all $\vec{x}$.

(c) Show that there is a positive integer constant k_M' (which depends on **M**) such that $\text{MA}time_M(\vec{x}) \leq (k_M')^{\text{MA}space_M(\vec{x})}$ for all $\vec{x}$.

We now have time and space measures for the RAM program, Turing machine, and Markov algorithm programming systems; we need some "natural" complexity measure for partial recursive function programs. Basically, we shall define partial recursive function "space" to be the maximum length of strings used in "computations" by partial recursive function programs. Specifically, for programs which simply specify one of the base functions we define $\text{PRF}space$ as follows: $\text{PRF}space_E(x) = |x|$, $\text{PRF}space_{S_j}(x) = |x|+1$, and $\text{PRF}space_{P_j^n}(\vec{x}) = |\vec{x}|$ for all x and $\vec{x}$. Suppose that **D** is a partial recursive function program which computes ϕ by substitution from $\theta_1, \ldots, \theta_m$ and ψ; then

$$\text{PRF}space_D(\vec{x}) = max\{\text{PRF}space_{\theta_1}(\vec{x}), \ldots,$$
$$\text{PRF}space_{\theta_m}(\vec{x}), \text{PRF}space_\psi(\theta_1(\vec{x}), \ldots, \theta_m(\vec{x}))\}$$

for all $\vec{x}$. The definitions of PRF*space* for recursion and minimization are the obvious variations on the preceding.

1.9.6 EXERCISE Define PRF*space* for definitions by recursion and minimization, and show that for any partial recursive function program **D** computing ϕ, $|\phi(\vec{x})| \leq$ PRF*space*$_D(\vec{x})$ for all $\vec{x}$.

Now that we have defined complexity functions for our programming systems we can begin to examine the complexities of translated programs in terms of the complexities of the original programs.

1.9.7 PROPOSITION *Let* **P** *be any RAM program and let* **T** *be the equivalent Turing machine produced by the translation in the proof of Theorem 1.7.3. There are positive integer constants* k_P *and* k_P' *depending on the program* **P** *such that*

$$\text{TM}space_T(\vec{x}) \leq \text{RAM}space_P(\vec{x}) + k_P$$

and

$$\text{TM}time_T(\vec{x}) \leq k_P'(\text{RAM}time_P(\vec{x}) + 1)^3$$

for all $\vec{x}$.

Proof Let the RAM program **P** use no registers other than R1, . . . , Rm. First, recall that the proof of Theorem 1.7.3 assumes that the RAM program uses instructions of types 1, 2, 6, and 7 only. Let **P'** be the RAM program equivalent to **P** using only these types of instructions which is produced by Proposition 1.6.7. We must relate the complexity of **P'** to that of **P**. Assuming that Exercise 1.6.8 has been worked in a reasonable way, **P'** uses no registers other than R1, . . . , Rm + 2 and

$$\text{RAM}space_{P'}(\vec{x}) \leq \text{RAM}space_P(\vec{x}) + 1.$$

1.9.8 EXERCISE Verify this last assertion.

The proof of Theorem 1.7.3 then completes the translation by producing a Turing machine **T** equivalent to **P'**. The space that **T** uses is exactly what is necessary to represent the contents of the RAM registers, and the length of this representation will be equal to the sum of the lengths of the RAM register contents plus the number of separating commas. Therefore,

$$\text{TM}space_T(\vec{x}) \leq \text{RAM}space_{P'}(\vec{x}) + m + 2$$

for all $\vec{x}$, and hence by the previous paragraph,

$$\text{TM}space_T(\vec{x}) \leq \text{RAM}space_P(\vec{x}) + m + 3.$$

Thus taking $k_P = m + 3$, completes the proof of the proposition with regard to space.

The analysis of the translation with regard to time is slightly more complicated. Again assuming that Exercise 1.6.8 has been worked in a reasonable way, a careful examination of the code used to replace instructions and produce the program $\mathbf{P}'$ shows that there is a constant k_1 such that for all $\vec{x}$,

$$\text{RAM}time_{P'}(\vec{x}) \le k_1 \text{RAM}time_P(\vec{x})\text{RAM}space_P(\vec{x}).$$

Now, examining the construction of the Turing machine $\mathbf{T}$ in the proof of Theorem 1.7.3 shows that there is a constant k_2 such that

$$\text{TM}time_T(\vec{x}) \le k_2 \text{RAM}time_{P'}(\vec{x})\text{TM}space_T(\vec{x})$$

for all $\vec{x}$; combining this with the inequalities established above and the observation after Exercise 1.9.2 yields that there are constants k_3 and k_P' such that

$$\text{TM}time_T(\vec{x}) \le k_3 \text{RAM}time_{P'}(\vec{x})[\text{RAM}space_{P'}(\vec{x}) + 1]$$
$$\le k_P'(\text{RAM}time_P(\vec{x}) + 1)^3$$

for all $\vec{x}$, which completes the proof of the proposition.   □

The analysis of the translation from Turing machines to Markov algorithms is straightforward, and we leave most of it as an exercise.

1.9.9 PROPOSITION   *Let* $\mathbf{T}$ *be any Turing machine and let* $\mathbf{M}$ *be the equivalent Markov algorithm produced by the translation in the proof of Theorem 1.8.3. Then*

$$\text{MA}space_M(\vec{x}) = \text{TM}space_T(\vec{x}) + 3$$

and

$$\text{MA}time_M(\vec{x}) \le \text{TM}time_T(\vec{x}) + 2\text{TM}space_T(\vec{x}) + 5 \le 3\text{TM}time_T(\vec{x}) + 5$$

for all $\vec{x}$.

1.9.10 EXERCISE   Prove the previous proposition.

We now turn our attention to the translation from Markov algorithms to partial recursive function programs.

1.9.11 PROPOSITION   *Let* $\mathbf{M}$ *be any Markov algorithm and* n *any positive integer, and let* $\mathbf{D}$ *be the partial recursive function program for the function of* n *arguments computed by* $\mathbf{M}$ *which is produced by the proof of Theorem 1.8.7. Then there is a positive integer constant* k

depending on **M** *and n such that*

$$\text{PRF}space_D(\vec{x}) \le k(\text{MA}time_M(\vec{x}) + 1)$$

for all $\vec{x}$.

Proof A careful examination of the proofs of Propositions 1.8.5 and 1.8.6 and the definition of the function I_n shows that all of the definitions **D'** of the (primitive recursive) functions and predicates involved have the following property: there is a constant k' depending on **M** or n such that if d' is the function defined by **D'** then

$$\text{PRF}space_{D'}(\vec{x}) \le max\{|\vec{x}|, |d'(\vec{x})|, \text{MA}space_M(\vec{x})\} + k'$$

for all $\vec{x}$; that is, the length of all strings in the computation of **D'** is bounded by the total length of the arguments, the length of the value, or the maximum length of a string in **M**'s computation, plus some constant (where $\text{MA}space_M(\vec{x})$ is needed only for M and $Stop_M$). Then from the final definition given in the proof of Theorem 1.8.7 we have that there is some constant k'' depending on **M** and n such that

$$\text{PRF}space_D(\vec{x}) \le max\{\text{MA}space_M(\vec{x}), \text{MA}time_M(\vec{x})\} + k''$$

for all $\vec{x}$. Finally, Exercise 1.9.5b yields that there is a constant k depending on **M** and n such that

$$\text{PRF}space_D(\vec{x}) \le k(\text{MA}time_M(\vec{x}) + 1)$$

for all $\vec{x}$.

1.9.12 EXERCISE   Supply the missing details from the proof of Proposition 1.9.11. $\square$

Finally, we complete our circle of translations by examining the translation from partial recursive function programs to equivalent RAM programs given in Section 1.6.

1.9.13 PROPOSITION   *Let* **D** *be any partial recursive function program and let* **P** *be the equivalent RAM program produced by the proof of Theorem 1.6.6. Then there is a constant k_D depending on* **D** *such that*

$$\text{RAM}space_P(\vec{x}) \le k_D\text{PRF}space_D(\vec{x})$$

for all $\vec{x}$.

1.9.14 EXERCISE   Prove the previous proposition.

The results of this section show that the translation of programs among the four programming systems from Sections 1.5–1.8 produces equivalent programs with only "modest" increases in complexity. Specifically, the space or time complexity of the translated program is bounded by the composition of a small degree polynomial with the space or time complexity (respectively) of the original program. In Chapter 3 we shall define a notion of "acceptable programming systems" which includes all "natural" programming systems, and shall show that any acceptable programming system can be translated effectively into any other. In Chapter 5 we shall define the notion of a general "computational complexity measure" on an acceptable programming system which includes all "natural" complexity measures on natural programming systems. We shall also show that, given any two acceptable programming systems and complexity measures on them, there is some total recursive function such that the complexity of translated programs can be bounded by the composition of this recursive function with the complexity of the original program. Because the notions of acceptable programming systems and general complexity measures allow some very "unnatural" systems and measures, this bounding result cannot be improved to show that the total recursive function is particularly "simple." However, the universal experience with "natural" programming systems is that the complexity of translated programs can be bounded by the composition of a very simple (usually small degree polynomial) recursive function with the complexity of the original program.

Additional Exercises

*1.9.15   Let **T** be any Turing machine and n any positive integer, and let **D** be the partial recursive function program for computing the function of n arguments produced by the translations in the proofs of Theorems 1.8.3 and 1.8.7. Show that there is a positive integer constant k_T depending on **T** such that

$$\text{PRF}space_D(\vec{x}) \leq 3\text{TM}time_T(\vec{x}) + k_T$$

for all $\vec{x}$. *Hint*: if **M** is the Markov algorithm from Theorem 1.8.3, then the constant k_M in Exercise 1.9.5 is equal to 1.

1.9.16   Give a direct translation from Markov algorithms to RAM programs and show that this translation yields RAM programs whose space and time complexity are each bounded by the composition of appropriate small degree polynomials with the space and time complexity, respectively, of the original Markov algorithms.

1.9.17   Define a "time" measure PRF*time* on partial recursive function programs and prove an appropriate version of Proposition 1.9.13 for RAM*time* and PRF*time*.

1.10   THE COMPLEXITY OF PRIMITIVE RECURSIVE AND ELEMENTARY FUNCTIONS

There are several senses in which the primitive recursive functions are "relatively simply computable." It is the goal of this section to prove results which give one of the precise and compelling interpretations of this intuition: namely that the class of primitive recursive functions is *closed* in the sense that a function is primitive recursive if and only if in some "natural" programming system under some "natural" complexity measure there is a program for computing it which has complexity which is itself bounded by a primitive recursive function. Based on these results, it is generally accepted that the class of primitive recursive functions (amply) includes all of the effectively computable functions which one might actually want to compute *in practice*. This is not to say that the primitive recursive functions are the only effectively computable functions of interest to "practical" computing. In fact the partial recursive functions which are not primitive recursive play an important role in the theory of specific algorithms and computational problems of great practical interest. Moreover, although there are "restricted" programming systems for computing just the primitive recursive functions (one such system was given when the primitive recursive functions were defined), such restricted programming systems suffer from some very serious disadvantages. For example, in Section 5.4 we prove that for any such restricted programming system there are functions whose shortest (restricted) programs are *enormously* longer than programs for computing them in general programming systems such as the partial recursive function programs. Therefore, although the class of primitive recursive functions includes all of the functions one might want to compute in practice, we can not always hope to simplify or strengthen our general theory of algorithms by restricting our attention to primitive recursive functions and programming systems which compute only them.

There are in fact small subclasses of the primitive recursive functions for which it is generally accepted that they also amply include all of the functions one might actually want to compute in practice. One such class is called the elementary functions, and our results in this section apply to it as well. Define the primitive recursive function $ex: N^2 \to N$ by $ex(0, n) = n$ and $ex(m + 1, n) = 2^{ex(m,n)}$ for all m and n; thus $ex(m, n)$ is given by a "stack" of m 2's topped off with an n. For an alphabet A_k, we say that a function $f: (A_k{}^*)^n \to A_k{}^*$ is bounded by a function $g: N \to N$ if for all $\bar{x}$, $|f(\bar{x})| \leq g(|\bar{x}|)$. For all m, all of the base functions for the primitive recursive functions are m-*elementary*, and a primitive recursive function f is m-*elementary* if f is obtained from m-elementary functions by substitution or recursion and if f is bounded by a function g such that $g(n) = ex(m, n) + k$ for some fixed k. In other words, a function f is m-elementary if there is a primitive recursive definition of f and all of the functions in the definition are bounded by some function g such that $g(n) = ex(m, n) + k$ for some constant k. Finally, we say that a function is *elementary* if it is m-elementary for some natural number m. Obviously every elementary function is primitive recursive.

All of the specific functions (except, of course, for ex) which we have shown so far in this book to be primitive recursive are in fact elementary. Actually, a careful (and tedious) check will verify that they are all m-elementary for a fairly small value of m, say 4 to be safe. This will also be the case for virtually all of the *specific* functions we show to be primitive recursive throughout the rest of this book. Thus, whenever a function is shown to be primitive recursive, the results of this section can usually be applied to conclude that it is "fairly simply computable" in the sense of being at most 4-elementary.

We establish the main result of this section with the following propositions.

1.10.1 PROPOSITION *Suppose a function f is computed by a program whose space or time complexity is bounded by a primitive recursive or elementary function in any one of the RAM, Turing machine, Markov algorithm, or partial recursive function programming systems. Then f has programs which compute it in space and time bounded by a primitive recursive or elementary function, respectively, in all of these programming systems.*

Proof The proof is immediate by the results of the previous section which show that any two of these "complexity measures" are related by at worst an exponential factor. □

1.10.2 PROPOSITION *Let f be a primitive recursive or elementary function; then f has programs which compute it in time and space bounded by a primitive recursive or elementary function, respectively, in all of the programming systems mentioned in the previous proposition.*

Proof Let f have a primitive recursive function program **D** for computing it which uses the functions $g_0 = f, g_1, \ldots, g_m$. For each g_i, define $\bar{g}_i : N \to N$ by

$$\bar{g}_i(n) = max\{|g_i(\vec{x})| : |\vec{x}| \le n\}$$

for all n. Then each $\bar{g}_i$ is primitive recursive, and if g_i is elementary then so is $\bar{g}_i$. Define $G : N \to N$ by

$$G(n) = max\{n, \bar{g}_0(n), \ldots, \bar{g}_m(n)\}$$

for all n. Then G is also primitive recursive, and if all of the g_i's are elementary then so is G. Finally, $\text{PRF}space_D(\vec{x}) \leq G(|\vec{x}|)$ for all $\vec{x}$ and since G is elementary if f is, the proof of the proposition is complete by applying Proposition 1.10.1.

1.10.3 EXERCISE   Fill in the missing details in the preceding proof. $\square$

1.10.4 PROPOSITION   *Suppose a function f is computed by a program in space or time bounded by a primitive recursive or elementary function in any of the programming systems mentioned in Proposition 1.10.1. Then f is primitive recursive or elementary, respectively.*

Proof   By proposition 1.10.1, there is a Markov algorithm **M** which computes f and a primitive recursive or elementary function G such that $\text{MA}time_M(\vec{x}) \leq G(|\vec{x}|)$ for all $\vec{x}$. Then for all $\vec{x}$,

$$|min_1 y[Stop_M(\vec{x}, y)]| \leq G(|\vec{x}|)$$

and therefore

$$f(\vec{x}) = \mathbf{M}(x) = M(\vec{x}, min_{y/G(|\vec{x}|)}[Stop_M(\vec{x}, y)])$$

(see Proposition 1.8.6 and the note following it). This shows that f is primitive recursive. Moreover, a straightforward check of the definitions of the functions and predicates in Propositions 1.8.5 and 1.8.6 shows that they are all elementary; therefore, if G is elementary then so is f.

1.10.5 EXERCISE   Verify the last sentence in the previous proof. $\square$

We summarize the results of the previous three propositions in the following theorem.

1.10.6 THEOREM   *Let the RAM, Turing machine, Markov algorithm, and partial recursive function programming systems be the programming systems under consideration.*

1. A function f is primitive recursive if and only if in one of the programming systems above there is a program which computes f in space or time bounded by a primitive recursive function; also, f is primitive recursive if and only if in each of the programming systems above there are programs which compute f in space and time bounded by a primitive recursive function.

2. A function f is elementary if and only if in one of the programming systems above there is a program which computes f in space or time bounded by an elementary function; also, f is elementary if and only if in each of the programming systems above there are programs which compute f in space and time bounded by an elementary function.

In Chapter 5 we shall discuss and clarify the *intuitive* notion of a "natural complexity measure" on a "natural programming system." *Then* this theorem can be interpreted as saying that for any natural complexity measure on a natural programming system, a function is primitive recursive or elementary if and only if there is a program which computes it whose complexity is bounded by a primitive recursive or elementary function, respectively. This is what we meant by saying that the primitive recursive and elementary functions are "relatively simply computable." It should also be clear that any function we might actually want to try to compute in practice is certainly elementary. (If a function could not be computed in time bounded by $ex(m, n) + k$ for *some* fixed m and k on inputs of length n, would *you* want to wait around for the output?).

Chapter 2
Reductions, Coding of RAM Programs, Unsolvable Problems

In addition to supplying evidence for the Church-Turing Thesis, the results of Chapter 1 also suggest that the class of partial recursive functions is extremely insensitive to the particulars of the programming system, or model of computation, which is used to define it. Thus, even from a very simple basis such as Turing machines, we get all of the partial recursive functions. One of our three main goals in this chapter is to show some simplifications of the basis for the partial recursive functions in a somewhat different direction. For example, we show that, for the sake of studying the partial recursive functions as we do later in this book, it is sufficient to consider only the partial recursive functions of one argument over the natural numbers. This simplification is useful for the further development of our theory.

Our second main goal in this chapter is to give an explicit indexing, or Gödel numbering, of the partial recursive functions based on RAM programs, and to prove some fundamental properties of this indexing. Such an indexing of the partial recursive functions by the natural numbers allows an identification between natural numbers and programs in such a way that programs can operate on natural number inputs as if they were programs; in other words, we develop the ability to have programs operate on other programs. In practice, this ability is usually obtained through the use of sophisticated string processing facilities in a programming language. Although such an approach is undoubtedly more natural and reasonable in practical situations, we take an alternative route here for two reasons. First, by developing this ability for a very simple programming system for the partial recursive functions of one argument over the natural numbers, we have this capability in a programming system for which it is comparatively easy to prove things *about* the system. Second, we hope to convince the reader by our example that this process for enabling programs to operate on other programs can be carried out for *any* programming system.

Finally, we give some important classical examples of algorithmically unsolvable problems. In addition to showing the unsolvability of the halting problem, in Sections 2.5 and 2.6 we show the unsolvability of the word problem for monoids (semigroups with identity) and the unsolvability of the Post Correspondence Problem. The first of these is probably of greatest interest to those with a strong mathematical bent; the second has many applications in theoretical computer science, particularly in formal language theory. While both of these sections are optional and are not essential for understanding later chapters, we recommend that all readers cover at least one of these sections, both to begin to get a feeling for some applications of the theory and also because these sections use ideas and techniques which we use again later in this book. Section 2.6 on the Post Correspondence Problem is especially relevant in this regard.

2.1 PAIRING FUNCTIONS AND REDUCTION
TO FUNCTIONS OF A SINGLE ARGUMENT

In this section we develop some useful correspondences between N^n and N. We use these for two purposes. First, given an effective one-to-one correspondence between the n-tuples in N^n and the numbers in N, we can reduce all computations over N involving functions of several arguments to ones involving only functions of a single argument. Second, with such a correspondence we can treat single numbers as finite sequences of numbers for the purposes of coding. We do this in later sections. We begin by reviewing some primitive recursive functions over A_1^* which were developed in Chapter 1; as in the exercises at the end of Section 1.4, we again identify A_1^* with N by identifying a_1^n with n. The following proposition merely summarizes results from Proposition 1.4.4 and Exercise 1.4.7 (in the context of functions over N) and results from Exercises 1.4.8 and 1.4.9.

2.1.1 PROPOSITION

1. *The following functions and predicates are primitive recursive over*
   N:

   (a) $n + m$　　　　　　　(b) $n \cdot m$　　(c) n^m
   (d) $n \dot{-} m = a_1^n \dot{-} a_1^m$　(e) $n \leq m$　(f) $n < m$.

   (Note that (d) defines $\dot{-}$ as the function $-$on strings from Exercise 1.3.7e.)

2. *If P is a primitive recursive predicate, then the following predicates and functions are also primitive recursive:*

   (a) $\exists m \leq nP$　(b) $\forall m \leq nP$　(c) $\min m \leq nP$　(d) $\max m \leq nP$,

   where we have replaced the "initial segment" notation "m/n" by "$m \leq n$", which is more natural in the context of our identification of A_1^* with N.

We now introduce the notion of a pairing function, from which we build our correspondences between N^n and N. A *pairing function* $\langle \, , \, \rangle$ is any primitive recursive function mapping N^2 to N which is both one-to-one and onto N and is also strictly monotone in each of its arguments; that is, $\langle x, y \rangle < \langle x + 1, y \rangle$ and $\langle x, y \rangle < \langle x, y + 1 \rangle$ for all x and y. (It follows that $x \leq \langle x, y \rangle$ and $y \leq \langle x, y \rangle$ for all x and y.) If $\langle \, , \, \rangle$ is a pairing function, then the *(primary) projection functions* Π_1 and Π_2 are defined by $\Pi_1(\langle x, y \rangle) = x$ and $\Pi_2(\langle x, y \rangle) = y$. Note that Π_1 and Π_2 are functions of a *single* argument, and they are also primitive recursive since

$$\Pi_1(z) = min \; x \leq z \, \exists y \leq z[\langle x, y \rangle = z]$$

and

$$\Pi_2(z) = min \; y \leq z \, \exists x \leq z[\langle x, y \rangle = z]$$

for all z. Also, it is obvious that $\langle \Pi_1(z), \Pi_2(z) \rangle = z$ for all z.

2.1.2 EXERCISE If $\langle \, , \, \rangle$ is any pairing function, what natural number is $\langle 0, 0 \rangle$?

For one example of a pairing function, note that since every positive integer can be factored uniquely into its even and odd "components," the function f such that $f(x, y) = 2^x(2y + 1)$ maps N^2 one-to-one and onto $N - \{0\}$. Thus $\langle \, , \, \rangle$ defined by $\langle x, y \rangle = 2^x(2y + 1) - 1$ is a pairing function.

2.1.3 EXERCISE Give fairly simple and direct descriptions of Π_1 and Π_2 for this definition of $\langle \, , \, \rangle$.

For another example of a pairing function, note that there are $n + 1$ pairs of integers (x, y) such that x is even and $(x/2) + y = n$ and there are n pairs of integers (x, y) such that x is odd and $((x + 1)/2) + y = n$. (For any expression z designating a real number, we also use the expression z to designate the greatest integer less than or equal to z. For example $z/2$ and $\sqrt{z}$ are always interpreted as integers.) Thus for each n, there are $2n + 1$ pairs (x, y) such that $((x + 1)/2) + y = n$ and in these $2n + 1$ pairs, x ranges from 0 through $2n$. Recall that $1 + 3 + 5 + \ldots + 2n - 1 = n^2$, so there are n^2 pairs (x, y) such that $((x + 1)/2) + y < n$. Therefore, for each $n \in N$ we can match the $2n + 1$ pairs (x, y) such that $((x + 1)/2) + y = n$ with the natural numbers

$$n^2 = (((x + 1)/2) + y)^2 + 0$$
$$n^2 + 1 = (((x + 1)/2) + y)^2 + 1$$
$$\vdots$$
$$n^2 + 2n = (((x + 1)/2) + y)^2 + 2n = (n + 1)^2 - 1$$

and get a one-to-one correspondence between N^2 and N. Thus $\langle \, , \, \rangle$ defined by $\langle x, y \rangle = (((x + 1)/2) + y)^2 + x$ is a pairing function with

$$\Pi_1(z) = z - (\sqrt{z})^2$$

and

$$\Pi_2(z) = \sqrt{z} - (\Pi_1(z) + 1)/2.$$

2.1.4 EXERCISE Verify that $\langle \, , \, \rangle$ as defined above is in fact primitive recursive, and that the formulas given for Π_1 and Π_2 are correct.

It is easy to see by the basic properties of one-to-one correspondences that if $\langle \, , \, \rangle$ is any pairing function, then $\langle \, , \, , \, \rangle_3$ defined by $\langle x, y, z \rangle_3 = \langle x, \langle y, z \rangle \rangle$ is a one-to-one correspondence between N^3 and N. And in general, for each $n > 2$ we define

$$\langle x_1, x_2, \ldots, x_{n-1}, x_n, x_{n+1} \rangle_{n+1} = \langle x_1, x_2, \ldots, x_{n-1}, \langle x_n, x_{n+1} \rangle \rangle_n,$$

giving us a one-to-one correspondence $\langle \, , \ldots, \, \rangle_n$ between N^n and N for each $n > 2$. Note that

$$\langle x_1, \ldots, x_{n-1}, x_n \rangle_n = \langle x_1, \langle \ldots, \langle x_{n-1}, x_n \rangle \ldots \rangle \rangle$$

and

$$\langle x_1, x_2, \ldots, x_{n+1} \rangle_{n+1} = \langle x_1, \langle x_2, \ldots, x_{n+1} \rangle_n \rangle.$$

Since the subscript n on $\langle \, , \ldots, \, \rangle_n$ is usually either clear from the context or irrelevant, we shall henceforth usually omit it. Notice that $\langle x_1, \ldots, x_n \rangle$ is always a *single* integer, z. We now define a primitive recursive (*uniform*) *projection function* Π such that if $n \geq 2$, $1 \leq i \leq n$, and $\langle x_1, \ldots, x_n \rangle = z$ then $\Pi(i, n, z) = x_i$; that is, if we wish to regard the single integer z as if it were an n-tuple then $\Pi(i, n, z)$ is the ith component of that n-tuple. Clearly, we do not *care* what $\Pi(i, n, z)$ is except when $n \geq 2$ and $1 \leq i \leq n$; so for all i and z we define

$$\Pi(i, 0, z) = 0$$
$$\Pi(i, 1, z) = z$$
$$\Pi(i, 2, z) = \Pi_1(z) \qquad \text{if} \quad i \leq 1$$
$$\Pi(i, 2, z) = \Pi_2(z) \qquad \text{if} \quad i > 1$$

and in general, from our definition of $\langle x_1, \ldots, x_n \rangle$ we see that a proper definition of Π for $n \geq 2$ is given by

$$\Pi(i, n + 1, z) = \begin{cases} \Pi(i, n, z) & \text{if} \quad i < n \\ \Pi_1(\Pi(n, n, z)) & \text{if} \quad i = n \\ \Pi_2(\Pi(n, n, z)) & \text{if} \quad i > n. \end{cases}$$

Π is primitive recursive by Exercise 1.4.11. Thus, starting with any pairing function we have constructed a *uniform* collection of primitive recursive one-to-one correspondences between N^n and N for $n \geq 2$.

2.1.5 EXERCISE

(a) For any x and n, show that

$$\langle 0, \ldots, 0 \rangle_n = 0 \quad \text{and} \quad \langle x, 0 \rangle = \langle x, 0, \ldots, 0 \rangle_n.$$

(b) For all n and z and all $i \geq n$, show that

$$\Pi(0, n, z) = \Pi(1, n, z) \quad \text{and} \quad \Pi(i, n, z) = \Pi(n, n, z).$$

(c) Define $\langle z \rangle_1 = z$. For all $n \geq 1$ and all z, show that

$$\langle \Pi(1, n, z), \ldots, \Pi(n, n, z) \rangle_n = z.$$

(d) For all i, n, and z, show that $\Pi(i, n, z) \leq z$.

(e) Show that there is a primitive recursive function *Large* such that

$$\Pi(i, n + 1, Large(n + 1, z)) = z$$

for all i, n, and z; we do not care what $Large(0, z)$ is. If you need a hint, see the end of the section.

The first use we make of our correspondences between N^n and N is to observe that *all* of the partial recursive functions of *any* number of arguments over N are "buried" in the partial recursive functions of *one* argument over N. Specifically, if ϕ^n is any partial recursive function of n arguments over N, then the partial function ϕ^1 of one argument over N defined by $\phi^1(z) = \phi^n(\Pi(1, n, z), \ldots, \Pi(n, n, z))$ is also partial recursive; also, if ϕ^1 is any partial recursive function of one argument over N then the corresponding partial function ϕ^n of n arguments over N defined by $\phi^n(x_1, \ldots, x_n) = \phi^1(\langle x_1, \ldots, x_n \rangle)$ is also partial recursive. Thus ϕ^1 is merely a coded representation of ϕ^n, and so, given the partial recursive functions of one argument over N together with a pairing function, we have access to all of the partial recursive functions over N.

2.1.6 EXERCISE
In the preceding discussion, ϕ^n is primitive recursive if and only if ϕ^1 is primitive recursive. In less than ten seconds, explain why.

Our effective reduction of all of the partial recursive functions over N to the partial recursive functions of one argument over N makes it possible to restrict our attention to just functions of one argument in the further development of our theory, and thereby to simplify the theory. Of course, it may sometimes be notationally convenient to write functions with more than one argument.

Additional Exercises

*2.1.5e $Large\,(n + 1, z) = \langle z, \ldots, z \rangle_{n+1}$.

*2.1.7 Define the function Con so that for all $m \geq 1$, $n \geq 1$, x, and y,

$$\Pi(i, m + n, Con(m, x, y)) = \begin{cases} \Pi(i, m, x) & \text{if } i \leq m \\ \Pi(i - m, n, y) & \text{if } i > m. \end{cases}$$

Show that Con is primitive recursive; also explain what Con does and why it does not need to have n as one of its arguments.

2.1.8 Show that the following two functions are primitive recursive.

(a) The function Rot such that for all $n \geq 1$ and all $x_1, \ldots, x_n$,

$$Rot(n, \langle x_1, x_2, \ldots, x_n \rangle) = \langle x_n, \ldots, x_2, x_1 \rangle.$$

(b) The function $Rotbeg$ such that for all $n \geq 1$ and all $x_1, \ldots, x_{n+1}$,

$$Rotbeg(n, \langle x_1, x_2, \ldots, x_n, x_{n+1} \rangle) = \langle x_n, \ldots, x_2, x_1, x_{n+1} \rangle.$$

2.1.9 Show how to write a RAM program, or "subroutine," over $A_1{}^*$ $(= N)$ for each $n \geq 1$ which if it is given $\langle n, x \rangle$ in register R1 finishes with $\Pi(i, n, x)$ in register Ri for all $1 \leq i \leq n$.

2.1.10

(a) Show that the function $\langle \, , \, \rangle$ defined by

$$\langle x, y \rangle = (x^2 + 2xy + y^2 + 3x + y)/2$$

is a pairing function.

(b) Give explicit definitions of Π_1 and Π_2 for this pairing function which do not refer to "$\langle \, , \, \rangle$."

2.1.11 Suppose that instead of defining $\langle \, , \ldots , \, \rangle_{n+1}$ as we did, we define $\langle x, y, z \rangle_3 = \langle \langle x, y \rangle, z \rangle$ and for $n \geq 3$

$$\langle x_1, \ldots, x_n, x_{n+1} \rangle_{n+1} = \langle \langle x_1, \ldots, x_n \rangle_n, x_{n+1} \rangle.$$

(a) What is the difficulty you encounter in proving that the appropriate projection function Π is primitive recursive for this definition of $\langle \, , \ldots , \, \rangle_n$?

(b) In spite of the difficulty, prove that Π is nevertheless primitive recursive.

*2.1.12 Define the primitive recursive function F by

$$F(x, y) = \Pi(y + 1, \Pi_1(x) + 1, \Pi_2(x)).$$

(a) Define the partial functions ψ_0, ψ_1, ψ_2, ... by

$$\psi_x(y) = \begin{cases} F(x, y) & \text{if} \quad y < \Pi_1(x) + 1 \\ undefined & \text{if} \quad y \geq \Pi_1(x) + 1. \end{cases}$$

Show that the sequence of functions ψ_0, ψ_1, ψ_2, ... contains each function which maps a nonempty finite initial segment $\{0, 1, \ldots, n\}$ of N into N. Also show that $\psi_x = \psi_y$ implies $x = y$.

(b) Define the partial functions π_0, π_1, π_2, ... by

$$\pi_x(y) = \begin{cases} F(x, y) - 1 & \text{if} \quad 0 < F(x, y) \text{ and } y < \Pi_1(x) + 1 \\ undefined & \text{otherwise.} \end{cases}$$

Show that the sequence of functions π_0, π_1, π_2, ... contains all functions which map finite subsets of N into N.

2.1.13 Show that the class of partial recursive functions of one argument over N can be obtained in a reasonable way without using any functions of more than one argument. That is, give a set of "base" functions of one argument and some "closure operations" on functions of one argument such that the class of partial recursive functions of one argument is the smallest class containing the base functions and closed under the closure operations.

2.1.14 Using the remark preceding Exercise 2.1.6 and using Exercise 2.1.13, show that the class of partial recursive functions over N can be obtained from only finitely many appropriately chosen "base" functions together with some appropriate "closure operations" on functions.

2.1.15 Let g, h, and k be primitive recursive functions, and let f be defined recursively by

$$f(0, x, y) = g(x, y) \quad \text{and} \quad f(z + 1, x, y) = f(z, h(x, y), k(x, y)).$$

Show that f is primitive recursive.

2.2 EQUIVALENCE OF ALPHABETS

In the preceding section we saw that as far as computations over N are concerned, it does not matter whether we calculate functions of several arguments or of just one argument, provided we do not mind the work of computing pairing and projection functions. Since we believe the Church-Turing Thesis, it is natural to suspect that it similarly does not matter over which *alphabet* we choose to compute. Of course, there is a trivial sense in which it does matter: A_1 simply does not have enough symbols to allow us to compute functions over, say, $A_{17}*$. What we expect is that, as was the case in the previous section, given some simple coding and decoding functions it does not matter which alphabet we use. In Section 1.4 we gave some simple coding and decoding functions, and in this section we show that given these, the partial recursive functions over all alphabets are "the same." Having done this, we are thereafter able to consider only partial recursive functions over $N = A_1*$ for the subsequent development of the theory.

In Section 1.4 we developed the primitive recursive coding and decoding functions C_k and D_k. C_k maps A_k* one-to-one and onto A_1*, and is given by

$$C_k(a_{i_1} \ldots a_{i_n}) = a_1^{\Sigma_{1 \le j \le n} i_j k^{n-j}};$$

C_k regards strings in A_k* as integers written to the base k. D_k is the inverse mapping (extended to all of A_k*) such that $C_k \circ D_k$ is the identity on A_1* and $D_k \circ C_k$ is the identity on A_k*. For any partial function ϕ over A_k* we define the partial function ϕ^+ over $A_1*(= N)$ by

$$\phi^+(x_1, \ldots, x_n) = C_k(\phi(D_k(x_1), \ldots, D_k(x_n))),$$

and for any partial function ψ over A_1* we define the partial function $\psi^\#$ over A_k* by

$$\psi^\#(x_1, \ldots, x_n) = D_k(\psi(C_k(x_1), \ldots, C_k(x_n))).$$

2.2.1 EXERCISE   What are $\phi^{+\#}$ and $\psi^{\#+}$?

It is a trivial matter to extend any partial function ψ over A_1* to one over A_k* by having $\psi(x) = \psi(a_1^{|x|})$, and the extended function is partial (or total, or primitive) recursive if the original one was.

2.2.2 EXERCISE   Prove the assertion we have just made.

Given this, it is then obvious that if ψ is partial (primitive) recursive, then so is $\psi^\#$. It may at first seem equally obvious that if ϕ is partial recursive, then so is ϕ^+, but in fact it is not so obvious. ϕ^+ was defined as the composition of partial recursive functions with ϕ, but these functions are over A_k*. What we need to do is to show how to define ϕ^+ *directly* over $A_1*(=N)$, without using any additional alphabet characters. This is our task for the remainder of this section.

First we need a couple of helpful primitive recursive functions over N. We want the function Q such that if the integer n is represented by the word wa_i in A_k* then $Q(n)$ is the integer represented by the word w. We could try $Q(n) = C_k(dell(D_k(n)))$ to show that Q is primitive recursive, but this is *cheating*; we have used functions over A_k*. However, as soon as we remember that multiplication by k "shifts digits" over one, we see that $Q(n) = \max\ m \leq n[k \cdot m < n]$ does work. Next we want the function R such that if n is represented by the word wa_i then $R(n)$ is i. Thus $R(n) = n \div k \cdot Q(n)$. Now our first step toward showing that if ϕ is partial recursive then so is ϕ^+ is to show that this is true for primitive recursive functions.

2.2.3 PROPOSITION   *Let f be a primitive recursive function over $A_k{}^*$, then f^+ over N is also primitive recursive.*

Proof   The proof is by induction on definitions of the primitive recursive functions. The case for the base functions is trivial: $E^+ = z$ (the zero function over N), $P_i{}^{n+} = P_i{}^n$, and $S_i{}^+(x) = kx + i$. The case for definitions by substitution is equally simple: suppose

$$f(x_1, \ldots, x_n) = g(h_1(x_1, \ldots, x_n), \ldots, h_m(x_1, \ldots, x_n)),$$

then

$$f^+(y_1, \ldots, y_n) = g^+(h_1{}^+(y_1, \ldots, y_n), \ldots, h_m{}^+(y_1, \ldots, y_n)).$$

The case for definitions by recursion is a bit tricky, however. Since we are dealing with codings of strings, if f is defined by recursion then the "next" value of f^+ depends not on the immediately "previous" value of f^+ but rather on some value of f^+ "farther back," as determined by the coding of words from $A_k{}^*$. To handle this more general type of recursive definition we introduce the function F such that $F(x)$ codes *all* values of f^+ up through $f^+(x)$.

To simplify notation, we assume that f is a function of only two arguments. Suppose that $f(\epsilon, y) = g(y)$ and $f(xa_i, y) = h_i(x, f(x, y), y)$ for $1 \leq i \leq k$. We want $f^+(0, y) = g^+(y)$, but we cannot simply define $f^+(x+1, y)$ directly from $f^+(x, y)$. Instead, $f^+(x+1, y)$ depends on the value of $f^+(Q(x+1), y)$. To handle this problem, we define the function F such that $F(x, y) = \langle f^+(0, y), \ldots, f^+(x, y) \rangle$, which yields

$$f^+(Q(x+1), y) = \Pi(Q(x+1) + 1, x+1, F(x, y)).$$

Since we also want

$$f^+(x+1, y) = h_i{}^+(Q(x+1), f^+(Q(x+1), y), y) \qquad \text{if} \quad R(x+1) = i$$

we define the primitive recursive function G by cases:

$$G(x, y, z) = h_i{}^+(Q(x+1), \Pi(Q(x+1) + 1, x+1, z), y) \qquad \text{if} \quad R(x+1) = i$$

for $1 \leq i \leq k$. Now we can define F by recursion with

$$F(0, y) = g^+(y)$$

and

$$F(x+1, y) = Con(x+1, F(x, y), G(x, y, F(x, y)))$$

where Con is from Exercise 2.1.7.

Finally, we see that f^+ is primitive recursive since

$$f^+(x, y) = \Pi(x+1, x+1, F(x, y)). \qquad \square$$

Note that the methods used in the proof of the previous proposition can be readily adapted to show that the primitive recursive functions are closed under what are called "course-of-values" recursions, that is recursions in which the next value of a function can depend on *all* previous values.

2.2.4 EXERCISE Explain why the proof of the previous proposition does not work for partial functions defined by recursion.

2.2.5 THEOREM *Let ϕ be a partial recursive function over $A_k{}^*$, then ϕ^+ over N is also partial recursive.*

Proof By the proof of Theorem 1.8.7 on Markov algorithms, there is a primitive recursive function f and a primitive recursive predicate P such that

$$\phi(x_1, \ldots, x_n) = f(x_1, \ldots, x_n, min_1 y P(y, x_1, \ldots, x_n)).$$

Define the helpful primitive recursive function c by $c(0) = 0$ and $c(z + 1) = c(z)k + 1$; thus $c(z) = C_k(a_1{}^z)$. Then we have

$$\phi^+(z_1, \ldots, z_n) = f^+(z_1, \ldots, z_n, c(min\ w P^+(c(w), z_1, \ldots, z_n)))$$

where the use of c enables us to search only over integers which code strings of a_1's. $\qquad \square$

Note that this proof, together with the previous proofs on which it depends, provides a method for converting an algorithm for computing ϕ into an algorithm for computing ϕ^+; it also provides a method (though not a very convenient one) for converting a definition of ϕ as a partial recursive function into a definition of ϕ^+ as a partial recursive function.

The previous theorem, together with the results of the preceding section, shows that through the use of some simple coding and decoding functions we can continue to study all of the partial recursive functions simply by studying only the partial recursive functions of one argument over N. We use this to simplify most of our further development in this book, though the simplifications are only in the realm of notation and terminology. No serious *conceptual* difficulties would arise in developing the theory for all partial recursive functions over any alphabet.

It may have occurred to you by now that this reduction to functions of a single argument over N is not entirely justified if we are thinking in terms of "computational complexity." Indeed, for very simple computable functions, the work of computing the coding and decoding functions may be far greater than that of computing the functions themselves. We are careful to take proper account of this observation at those later points when it is relevant.

2.3 REMOVAL OF PRIMITIVE RECURSIONS

In order to simplify our proofs of the Gödel incompleteness and undecidability results in Chapter 4 and of the intractability of the theory of addition in Section 6.4, we need yet another characterization of the partial recursive functions in addition to those provided by Chapter 1. In this (optional) section we give this characterization, which is over the natural numbers N and uses some carefully chosen base functions along with the operations of substitution and minimization. In addition to being used in Chapters 4 and 6, this characterization and the methods used for obtaining it may be of interest in their own right; in particular, they give a method for removing primitive recursions in favor of "simpler" operations.

A key property of a function defined by primitive recursion is that each value has a given relationship to "previous" values. If you think about it a few minutes, you might guess that a function defined by recursion could be obtained without explicitly using recursion if we could find and decode finite sequences of values which have the proper relationship among their members. Thus an obvious strategy for obtaining our characterization is to remove primitive recursions used to define partial functions ϕ by somehow using minimization to encode a sequence of values of the form $\phi(\epsilon)$, $\phi(a)$, $\phi(ab)$, . . . , $\phi(x)$. We faced this same problem in the last section where for a function f we encoded $f(0)$, . . . , $f(x)$ as $\langle f(0), . . . , f(x) \rangle$. Unfortunately, we used recursion several times in defining this coding and the corresponding projection function. We know of no *easy* way to define the coding and projection functions without using recursion, although the results of this section show that there is *some* way to do so.

Thus, our main technical task in this section is to develop a method for describing finite sequences which does not depend on recursion. Gödel accomplished this task by using the function β such that $\beta(x, y, z)$ is the remainder when $(x + 1)y + 1$ is divided by z. This function has the property that for any sequence of integers w_0, . . . , w_k there exist m and n such that $\beta(i, m, n) = w_i$ for $1 \leq i \leq k$. Although β is fairly easy to define, it is a bit tricky to prove that it has this nice property; the proof requires more number theory (including the Chinese Remainder Theorem) than we wish to assume or develop in this book. So we use a different coding scheme which uses the coding and decoding functions C_k and D_k from Section 1.4 to code strings and sequences of strings over *arbitrary alphabets* as natural numbers. As a result, we not only remove primitive recursions, but we also provide another proof of Theorem 2.2.5, showing that all partial recursive functions over arbitrary alphabets can be coded as partial recursive functions over N.

The characterization of the partial recursive functions over N which we now want is that they are the class of functions obtainable from addition, $+$, multiplication, $\cdot$, the projection functions, $P_i{}^n$, and the characteristic function of the equality predicate, $c_=$, by the operations of substitution and minimization. We refer to functions and predicates obtainable from addition, multiplication, the projection functions, and the equality predicate by substitution and minimization as being *min-computable*. Since addition, multiplication, the projection functions, and the equality predicate (that is, the base functions for the min-computable functions) are all primitive recursive, it is obvious that all of the min-computable functions are partial recursive. We break the proof of the converse into the next three propositions. The first provides some generally useful min-computable functions and predicates.

2.3.1 PROPOSITION

1. *The following functions and predicates are min-computable: all constant functions; $x \overset{\cdot}{-} y$; $x \bmod y$ (the remainder when x is divided by y); $x \div y$ (the quotient when x is divided by y); $x \le y$; and $x \mid y$ (x divides y evenly).*
2. *The min-computable predicates are closed under the Boolean operations of "not," "and," "or," and "implies," and they are closed under bounded quantification.*
3. *The min-computable functions are closed under definition by cases.*

Proof The following sequence of formulas proves the proposition, providing some extra min-computable functions in the bargain:

$$zero(x) = c_=(x, \min y[c_=(x, y) = 0]);$$
$$one(x) = c_=(x, x);$$
$$two(x) = one(x) + one(x)$$
$$three(x) = two(x) + one(x)$$
$$\vdots$$

$$
\begin{aligned}
1 \doteq x &= c_=(x, 0); \\
lev(x) &= 1 \doteq (1 \doteq x); \\
c_{not\,P} &= 1 \doteq c_P; \\
c_{P\,or\,Q} &= lev(c_P + c_Q); \\
c_{P\,and\,Q} &= c_{not(notP\,or\,notQ)}; \\
c_{P\,implies\,Q} &= c_{notP\,or\,Q}; \\
abs(x - y) &= min\ z[x + z = y\ or\ y + z = x]; \\
x \le y\ &iff\ x + abs(x - y) = y; \\
x \doteq y &= min\ z[x \le y + z]; \\
\exists x \le yP(x, \vec{z})\ &iff\ (min\ w[P(w, \vec{z})\ or\ w = y + 1]) \le y; \\
\forall x \le yP(x, \vec{z})\ &iff\ not\ \exists x \le y(not\ P(x, \vec{z})); \\
x \,|\, y\ &iff\ \exists z \le y(z \cdot x = y); \\
x\ mod\ y &= min\ z[\exists w \le x(w \cdot y + z = x)]; \\
x \div y &= min\ z[x = z \cdot y + (x\ mod\ y)];
\end{aligned}
$$

$$
\phi(\vec{x}) \cdot c_P(\vec{x}) + \psi(\vec{x}) \cdot c_{notP}(\vec{x}) = \begin{cases} \phi(\vec{x}) & \text{if } P(\vec{x}) \\ \psi(\vec{x}) & \text{otherwise.} \end{cases} \quad \square
$$

The next proposition provides some min-computable functions and predicates for simulating some basic string manipulations relating to the coding and decoding functions C_k and D_k (for "nice" values of k). Recall that $C_k : A_k^* \to N$ is given by

$$
C_k(a_{i_1} \ldots a_{i_n}) = \Sigma_{1 \le j \le n} i_j \cdot k^{n-j}.
$$

Multiplication by powers of k performs "shifts" in this base k representation of the integers; specifically, if u and v are strings in A_k^* then

$$
C_k(uv) = C_k(u) \cdot k^{|v|} + C_k(v).
$$

This shows what numerical manipulations are needed to perform concatenation of coded strings; getting $k^{|v|}$ is the key problem since we have not yet shown that exponentiation is min-computable.

2.3.2 PROPOSITION *Let p be a prime number, m a natural number, and $k = p^m$. The following functions and predicates are min-computable:*

1. $con(x, y) = C_k(D_k(x)D_k(y))$;
2. x/y iff $D_k(x)$ is an initial segment of $D_k(y)$;
3. $x \backslash y$ iff $D_k(x)$ is a final segment of $D_k(y)$; and
4. $x \lessgtr y$ iff $D_k(x)$ is a substring of $D_k(y)$.

(*Strictly speaking, each of these functions and predicates should be subscripted with k.*)

Proof First notice that it is easy to use min-computable functions and predicates to say that a number is a power of p or is a power of k:

$Power_p(x)$ iff $x \neq 0$ and $\forall y \leq x[(y \neq 1 \ and \ y|x) \ implies \ p|y]$;

$Power_k(x)$ iff $\exists y \leq x[Power_p(y) \ and \ y^m = x]$

where, since m is a fixed integer, y^m is an *abbreviation* for the m-fold product of y with itself and does not stand for exponentiation. Notice that the second predicate uses the fact that $(p^m)^n = (p^n)^m$.

Next notice that if the $n+1$-st symbol from the right end of $D_k(x)$ is a_j then $(x \div k^n) \ mod \ k$ is j (if $j \neq k$, otherwise it is 0). Thus we can say that x codes a string of a_1's as follows:

$D_k(x) \in A_1^*$ iff $x = 0$ or

$\forall y \leq x[Power_k(y) \quad implies \quad (x \div y) \ mod \ k = 1]$.

In the order imposed by C_k on the strings in A_k^*, a_1^n is the first string of length n and a_k^n is the last string of length n; moreover, $k \cdot C_k(a_1^n) = C_k(a_k^n)$. Thus we can say that two numbers code strings of equal lengths as follows:

$|D_k(x)| = |D_k(y)|$ iff $\exists z \leq x[D_k(z) \in A_1^* \quad and \quad z \leq x, y \leq k \cdot z]$.

Finally, notice that

$$k^{n+1} = \Sigma_{1 \leq j \leq n}(k - 1)k^j + k = C_k(a_{k-1}^n a_k).$$

Thus $D_k(k^{n+1}) = a_{k-1}^n a_k$ and so $|D_k(k^{n+1})| = n+1$. Thus we can obtain $k^{|D_k(x)|}$ as follows:

$$k^{|D_k(x)|} = min \ y[Power_k(y) \quad and \quad |D_k(y)| = |D_k(x)|].$$

Now, it is simple to define the concatenation function and the predicates which are required by the proposition:

$con(x, y) \quad = \quad x \cdot k^{|D_k(y)|} + y$;

x/y iff $\exists z \leq y[con(x, z) = y]$;

$x \backslash y$ iff $\exists z \leq y[con(z, x) = y]$;

$x \leqslant y$ iff $\exists w, z \leq y[con(w, con(x, z)) = y]$. $\square$

The previous proposition provides some min-computable functions and predicates for using integers as if they were strings in A_k^*, the strings which they code. To attain our goal in this section we shall also have to consider A_k as embedded in a somewhat larger alphabet A_n which contains some extra symbols which can be used as markers to separate strings in A_k^*. The next proposition shows that we can transform integers from base k to base n, and *vice versa*.

2.3.3 PROPOSITION *Let p be a prime number, m be a natural* *number, $k = p^m$, and $n = p^{m+1}$. Then the two functions $C_n \circ D_k$ and $C_k \circ D_n$ are each min-computable.*

Proof First we give a predicate which says that two numbers x and y are the same power q of k and n, respectively; that is, $x = (p^m)^q = (p^q)^m$ and $y = (p^{m+1})^q = (p^q)^{m+1}$:

$Samepower_{k,n}(x, y)$ iff $\exists w \leq x[Power_p(w)$ and $w^m = x$ and $w^{m+1} = y]$.

(Of course, w^m and w^{m+1} are abbreviations as in the previous proof.) Next

$$|D_k(x)| = |D_n(y)| \quad \text{iff} \quad Samepower_{k,n}(k^{|D_k(x)|}, n^{|D_n(y)|}).$$

Finally, since $C_n(u)$ uses n^q for just those q less than $|u|$,

$$D_k(x) = D_n(y) \quad \text{iff}$$

$|D_k(x)| = |D_n(y)|$ *and* $[x = y = 0$ *or* $\forall w, z \leq n^{|D_n(y)|} \div 1$
$[Samepower_{k,n}(w, z)$ *implies* $(x \div w) \bmod k = (y \div z) \bmod n]]$.

Now it is simple to define the functions we want:

$$C_n \circ D_k(x) = \min y[D_k(x) = D_n(y)];$$

$$C_k \circ D_n(x) = \begin{cases} \min y[D_n(x) = D_k(y)] & \text{if} \quad D_n(x) \in A_k{}^* \\ 0 & \text{otherwise.} \end{cases}$$

We leave showing that the predicate $D_n(x) \in A_k{}^*$ is min-computable as an exercise.

2.3.4 EXERCISE Show that $D_n(x) \in A_k{}^*$ is min-computable. ☐

The previous three propositions give the min-computable functions for string manipulations which are needed to complete our characterization of the partial recursive functions as the min-computable functions.

2.3.5 THEOREM *Let k be any positive integer, and for any partial recursive function ϕ over $A_k{}^*$ let ϕ^+ over N be defined by*

$$\phi^+(x_1, \ldots, x_n) = C_k(\phi(D_k(x_1), \ldots, D_k(x_n))).$$

Then ϕ^+ is min-computable.

Proof First we prove the theorem for those values of k which are powers of prime numbers; suppose that p is prime and $k = p^m$. The proof is by induction on the definitions of the partial recursive functions. The cases for base functions and for functions obtained by substitution are trivial (as in the proof of Proposition 2.2.3). The case for definitions by minimization is quite straightforward. Suppose that

$$\phi(\vec{x}) = min_j y[\psi(y, \vec{x}) = \epsilon]$$

then

$$\phi^+(\vec{z}) = min \ w[D_k(w) \in \{a_j\}^* \ and \ \psi^+(w, \vec{z}) = 0];$$

we leave as an exercise to show that $D_k(w) \in \{a_j\}^*$ is a min-computable predicate. The case for functions defined by primitive recursion is the heart of the matter.

Suppose that ϕ is defined by primitive recursion from ψ, $\theta_1, \ldots, \theta_k$. For *notational* convenience, we assume that ϕ has two arguments. We let $n = p^{m+1}$ and assume that $A_n - A_k$ contains the two new symbols # and \$ which we can use as markers. We shall define a min-computable function ϕ_{comp} which codes "computation" strings of the form

$$\#\epsilon\$\phi(\epsilon, y)\#a\$\phi(a, y)\# \ldots \#x\$\phi(x, y)\#$$

where a is the first letter in x. In order to do this, it is necessary to convert integers from base k to base n (so that we can use the markers), and then back again.

Since we are thinking of natural numbers as coding strings in $A_n{}^*$, we shall write them as if they were the strings they code; for example, $x\#y$ instead of $con(D_n(x), con(k + 1, D_n(y)))$ if $\# = a_{k+1}$. We define a helpful min-computable function ϕ_{next} such that for the string $\#v\$w\#$, if w is $\phi(v, y)$ then $\phi_{next}(i, v, w, y)$ codes the string $va_i\$\phi(va_i, y)\#$. For $1 \leq i \leq k$ and for v and w coding strings in $A_k{}^*$ *to the base n*,

$$\phi_{next}(i, v, w, y) = \begin{cases} va_1\$C_n \circ D_k(\theta_1^+(C_k \circ D_n(v), C_k \circ D_n(w), y))\# & if \ i = 1 \\ \vdots & \vdots \\ va_k\$C_n \circ D_k(\theta_k^+(C_k \circ D_n(v), C_k \circ D_n(w), y))\# & if \ i = k; \end{cases}$$

ϕ_{next} has the value 0 otherwise. (Note that since y never appears in the computation string, it does not have to undergo the base change.)

ϕ_{comp} is defined by

$$\phi_{comp}(x, y) = min \ z[\#\$C_n \circ D_k(\varphi^+(0, y))/z \quad and \quad \forall v \leq C_n \circ D_k(x) \forall i \leq k$$
$$(i \neq 0 \quad and \quad va_i / C_n \circ D_k(x) \quad implies$$
$$\exists w \leq z[\#v\$w\#\phi_{next}(i, v, w, y) \preccurlyeq z])];$$

taking z to be minimal is crucial, since there are longer strings which satisfy the conditions by starting correctly and then restarting incorrectly. Finally,

$$\phi^+(x, y) = C_k \circ D_n(min \ z[\$z\#\backslash \ \phi_{comp}(x, y)]).$$

Now that we have proved the theorem for prime power values of k, we have that all partial recursive functions over N are min-computable. In particular, we have that exponentiation is min-computable. This then makes adapting the proofs of Propositions 2.3.2 and 2.3.3 to any value of k and any $n \geq k + 2$ completely straightforward, and then the proof we have just given goes through for all k.

2.3.6 EXERCISE   Complete the proof of the previous theorem; the key to the part for arbitrary k and n involves the predicates $Power_k$ and $Samepower_{k,n}$.   □

The previous theorem gives the characterization of the partial recursive functions we need for Chapter 4. However, for Section 6.4 we need one other bit of information. The proof of the previous theorem actually gives a translation of any partial recursive function program into an equivalent "MIN-program," and we would like to know how the complexity of the resulting "MIN-program" relates to the complexity of the original partial recursive function program.

2.3.7 EXERCISE   Give a precise definition of the MIN-programs by making a slight variation in the definition of the partial recursive function programs. Also give a precise definition of MIN*space* functions such that if C is a MIN-program then MIN$space_C(\vec{x})$ is the largest integer used in C's computation on inputs $\vec{x}$.

First notice that *any* translation from programs over alphabet A_k with k greater than 1 to equivalent programs over $A_1{}^* = N$ *must* introduce an exponential increase in (space) complexity simply because of the exponential increase in the length of numbers when going from representation in base k to representation in unary. The next proposition shows that our translation does not do "too much" worse.

2.3.8 PROPOSITION   *Let* **D** *be any partial recursive function program, and let* **C** *be the MIN-program equivalent to* **D** *produced by the translation from the proof of Theorem 2.3.5. Then there is a positive integer constant c depending on the alphabet for* **D** *such that for all* $\vec{x}$

$$\text{MIN} space_C(\vec{x}) \leq 2^{c[(\text{PRF} space_D(\vec{x}))^2 + 1]}.$$

(Notice that we have deliberately used the same notation, $\vec{x}$, for the inputs to **D** and for the coded inputs to **C**, since we really want to think of **C** as working on integers written to some base k greater than 1.)

Proof   A careful examination of the proofs of Propositions 2.3.1, 2.3.2, and 2.3.3 and Theorem 2.3.5 shows that the only possible source of (coded) strings significantly longer than PRF$space_D(\vec{x})$ is in the removal of primitive recursions; specifically from the function ϕ_{comp} in the proof of Theorem 2.3.5. But for some (small) constant d we have that

$$|D_k(\phi_{comp}(\bar{x}))| \le d[|\bar{x}| \cdot \text{PRF}space_D(\bar{x}) + 1] \le d[(\text{PRF}space_D(\bar{x}))^2 + 1].$$

Then converting from strings to natural numbers yields that for some positive integer constant c

$$\text{MIN}space_C(\bar{x}) \le 2^{c[(\text{PRF}space_D(\bar{x}))^2 + 1]}.$$

2.3.9 EXERCISE   Fill in the details for the proof of the previous proposition.   ☐

For our purposes in Section 6.4 it is more convenient to have the complexity relationship given in the following theorem:

2.3.10 THEOREM   *Let* **T** *be any Turing machine and n be any positive integer, and let* **C** *be the MIN-program of n arguments produced by the translations in the proofs of Theorems 1.8.3, 1.8.7, and 2.3.5. Then there is a positive integer constant c depending on* **T** *such that for all* $\bar{x}$.

$$\text{MIN}space_C(\bar{x}) \le 2^{c[(\text{TM}time_T(\bar{x}))^2 + 1]}.$$

Proof   The proof uses Exercise 1.9.15 and Proposition 2.3.8.

2.3.11 EXERCISE   Complete the proof of the previous theorem.   ☐

Additional Exercises

2.3.12   Show that Gödel's function β has the nice property we described, and use it to give a "simpler" proof that all partial recursive functions over N are min-computable.

2.3.13   Let C be a MIN-program. Suppose we define $\text{MIN}length_C(\bar{x})$ to be the length of the largest integer used in C's computation on inputs $\bar{x}$ (where the integers are written to base $k > 1$), show that if we replace MIN*space* by MIN*length* in Proposition 2.3.8 and Theorem 2.3.10 that the exponential bounds reduce to small polynomial bounds. The reasons for using MIN*space* rather than MIN*length* will be clarified in Section 6.4.

2.3.14   Prove that *zero(x)* is min-computable without using minimization.

2.4   CODING OF RAM PROGRAMS

On the basis of the results of Sections 2.1 and 2.2, we now restrict our attention to the partial recursive functions over N, and in most cases to functions of a single argument. In this section we do some coding (sometimes called "arithmetization") of RAM programs (over N) in order that we may talk about RAM programs by talking about integers. Since the coding identifies programs with integers, we can use this coding to have our RAM programs "talk" about other RAM programs. The coding of RAM programs gives us what is sometimes called an indexing, or Gödel numbering, of the partial recursive functions. It turns out that "nice" Gödel numberings (those with certain fundamental properties) can be used to characterize all "acceptable" programming systems. We shall develop this characterization in Chapter 3.

Since we are working over N and since we have shown in Proposition 1.6.7 that not all of the RAM instructions are necessary, it is sufficient to consider RAM programs composed of the following five types of instructions, with each type of instruction coded as indicated:

1.	Ni	**add**	Rj		coded	$\langle 1, i, j, 0 \rangle$
2.	Ni	**del**	Rj		coded	$\langle 2, i, j, 0 \rangle$
3.	Ni	**continue**			coded	$\langle 3, i, 1, 0 \rangle$
4.	Ni	Rj	**jmp**	Nka	coded	$\langle 4, i, j, k \rangle$
5.	Ni	Rj	**jmp**	Nkb	coded	$\langle 5, i, j, k \rangle$

If your favorite RAM program does not have line names for each instruction, simply pick a handy, unused line name and apply it to all unnamed instructions. If I is a RAM instruction of one of these types, 'I' denotes the integer by which it is coded.

At this point you may wish to review Section 2.1, and in particular to refresh your familiarity with $\Pi(i, m, x)$ as the ith component of the integer x when it is viewed as an ordered m-tuple of integers. If 'I' is the code of an instruction as given above, then we can use the projection function to decode 'I' to determine the type (as given above) of I, the line name in I, the register changed or tested by I, and the line name to which I jumps if a jump is executed. For the sake of more readable notation, we introduce the simple primitive recursive functions Typ, Nam, Reg, and Jmp given by

$$Typ(x) = \Pi(1, 4, x), \qquad Nam(x) = \Pi(2, 4, x),$$
$$Reg(x) = \Pi(3, 4, x), \qquad Jmp(x) = \Pi(4, 4, x);$$

these functions yield the type, line name, register mentioned, and line name jumped to, respectively, for an instruction coded by x. Note that if x does not code an instruction then we have no need to interpret the values these functions happen to yield. In fact, generally for the rest of this section we have no need to interpret the values certain functions yield outside of their intuitively clear "intended domains"; the intended domain for the functions we have just given is the set of those numbers which happen to code an instruction.

As an example of how our coding and decoding of RAM instructions works, we now show that the predicate *INST* such that *INST(x)* if and only if *x* codes an instruction is primitive recursive. First we introduce the helpful Boolean combination *implies* and show that it preserves primitive recursive predicates by defining

$$P \text{ implies } Q \quad \text{iff} \quad (notP) \text{ or } Q.$$

Then we have that *INST(x)* holds if and only if

$$1 \le Typ(x) \le 5 \quad and \quad 1 \le Reg(x) \quad\quad and$$
$$[Typ(x) \le 3 \quad implies \quad Jmp(x) = 0] \quad\quad and$$
$$[Typ(x) = 3 \quad implies \quad Reg(x) = 1],$$

You should be able to verify easily that this assertion is correct.

We now continue by showing how to code and decode RAM programs. Let **P** be a RAM program composed only of instructions of the five types given above; say **P** $= I_1, \ldots, I_n$. We let '**P**' stand for the integer which codes **P**, where the coding of **P** is given by

$$\text{'}\mathbf{P}\text{'} = \langle n, \text{'}I_1\text{'}, \ldots, \text{'}I_n\text{'} \rangle = \langle n, \langle \text{'}I_1\text{'}, \ldots, \text{'}I_n\text{'} \rangle \rangle.$$

Note that the second equality above follows simply from the definitions of $\langle \, , \ldots , \rangle$ in Section 2.1, but it is quite important. Given '**P**' we want to be able to decode it and reconstruct **P**, and to begin that process we need to know how many instructions there are in **P**. The second equality above shows us how to get that information very easily. $\Pi(1, 2, \text{'}\mathbf{P}\text{'})$ is the number of instructions in **P**, that is the "length" of **P**, and $\Pi(2, 2, \text{'}\mathbf{P}\text{'})$ gives the "program" part of '**P**'. Again, for the sake of readability we define the simple primitive recursive functions *Ln*, *Pg*, and *Line* such that $Ln(x) = \Pi(1, 2, x)$, $Pg(x) = \Pi(2, 2, x)$, and $Line(i, x) = \Pi(i, Ln(x), Pg(x))$; these functions yield the length (number of instructions), program part, and code of the *i*th line of the program coded by *x*, respectively. Of course, if *x* does not code a program we have no need to interpret these functions. However it is interesting to note that because of the particular way in which we defined the uniform projection function Π, it happens that $Line(0, x) = Line(1, x)$ and $Line(i, x) = Line(Ln(x), x)$ for all $i \ge Ln(x)$ (see Exercise 2.1.5b).

As an example of how our coding and decoding of RAM programs works, we now show that the predicate *PROG* such that *PROG(x)* if and only if *x* codes a program is primitive recursive. A number codes a program just in case each "line" codes an instruction, each jump instruction has a place to jump to, and the last instruction is a **continue**. Thus $PROG(x)$ holds if and only if

$$\forall i \le Ln(x) \ (i \ge 1 \quad implies$$

$$[INST(Line(i, x)) \quad and \quad Typ(Line(Ln(x), \ x))=3$$

$$and \quad (Typ(Line(i, x))=4 \quad implies$$

$$\exists j \le i \dot{-} 1 [j \ge 1 \quad and \quad Nam(Line(j, x))=Jmp(Line(i, x))])$$

$$and \quad (Typ(Line(i, x))=5 \quad implies$$

$$\exists j \le Ln(x)[j>i \quad and \quad Nam(Line(j, x))=Jmp(Line(i, x))])]).$$

Note that we have used expressions such as $\exists x \le f(y)P(x)$ as introduced in Exercise 1.4.13.

2.4.1 EXERCISE

(a) Show that the function Clr is primitive recursive, where $Clr(i)$ is the code of the following RAM program which clears register Ri:

        N1   **del** Ri

        N1   Ri **jmp** N1a

        N1   **continue**;

(b) Show that the function $Conprogs$ is primitive recursive where $Conprogs(x, y)$ is the code of the program gotten by concatenating the program coded by y to the end of the program coded by x, assuming that x and y both code programs;

(c) Show that the function $Cumclr$ is primitive recursive where if $i \ge 2$ then $Cumclr(i)$ is the code of a program which clears registers R2, . . . , Ri.

Our coding of RAM programs allows us to use integers as if they were programs, and hence to have RAM programs operate on integers as if they were programs. We are going to use our coding and decoding of RAM programs to give primitive recursive functions which describe in detail the computations of RAM programs, but before we do this we present a fundamental result in the theory of algorithms known as the unsolvability of the halting problem.

2.4.2 THEOREM (Unsolvability of the Halting Problem) *There is no RAM program* **P** *such that if* **P** *starts with x in register R1 and with i in R2 (the other registers empty) then (1)* **P** *eventually halts, (2)* **P** *halts with 1 in R1 if i codes a program which eventually halts when started with x in R1 (and the other registers empty), and (3)* **P** *halts with 0 in R1 if i* *codes a program which does not halt when started with x in R1 (and the other registers empty).*

Proof Assume that **P** is such a RAM program, and let **Q** be the following RAM program:

$$R2 \leftarrow R1$$
$$\textbf{P}$$

N1   **continue**
         R1   **jmp**   N1a
         continue.

Then there is a RAM program of the restricted type we are considering in this section which is equivalent to **Q**; let q be the code of such a program. We are not interested in actually calculating q, although it is not hard to do, but we are instead interested in what happens when program **Q** is started with q in R1 and the other registers empty. When this is done we get **P** started with q in both R1 and R2. Because of the properties we have assumed for **P**, **P** eventually halts with either 1 or 0 in R1. If **P** halts with 1 in R1 then **Q** goes into an infinite loop, while if **P** halts with 0 in R1 then **Q** halts also. But again, by the properties we have assumed for **P**, if **P** halts with 1 in R1 then **Q** must halt on input q, and if **P** halts with 0 in R1 then **Q** does not halt on input q. Therefore we have that if **Q** is started with q in R1 then it halts if and only if it does not halt. This contradiction shows that no such program **P** can exist. If you are not absolutely sure you completely understand the argument we have just given, go back over it until you do.   □

Using the Church-Turing Thesis, the preceding theorem shows that there is no algorithm for determining whether or not arbitrary RAM programs halt. It also follows that there is no algorithm for determining whether Turing machines, Markov algorithms, etc., halt since there are algorithms for translating RAM programs into all such formalisms. To see this suppose, for example, that we had an algorithm for testing whether Turing machines halt. Then we could first apply our algorithm for translating RAM programs into Turing machines and then apply the supposed algorithm to the resulting Turing machines. We would thereby be testing whether the original RAM programs halt, and we know that cannot be done. This method of translating, or "reducing," the halting problem (for RAM programs) to other problems is a fundamental technique for showing that other problems have no algorithmic solution. There are many other examples of this later in this book, for example in Sections 2.5 and 2.6, and the general method is fundamental to understanding Chapter 7. The unsolvability of the halting problem for other systems such as Turing machines can also be established through the same sort of constructions and arguments that we have used for RAM programs. We do not go to that trouble in this book, but rather content ourselves with the argument based on the Church-Turing Thesis which we have just given, together with some very general results in Chapter 3.

We now return to developing the primitive recursive functions which describe the computations of RAM programs. Suppose that at some point $r1, \ldots, rn$ are the contents of the RAM registers $R1, \ldots, Rn$; we can code these into the single integer $\langle r1, \ldots, rn \rangle$. Conversely, given any integer x, we can view it as coding the contents of $R1, \ldots, Rn$ by taking the sequence $\Pi(1, n, x), \ldots, \Pi(n, n, x)$; but notice that to do so we need to know n, the number of registers we wish to "fill." If we are interested in the computations of RAM programs, then for any given program we only need to keep track of the contents of the registers actually mentioned in the program. From Exercise 2.1.5d we have that

$$Reg(Line(i, x)) \leq Line(i, x) \leq Pg(x) \leq x$$

for all i and x, and therefore if x codes a program, $R1, \ldots, Rx$ certainly include all of the registers mentioned in the program. Thus if we want to interpret any given integer y as coding the contents of the registers used by the program coded by x, $r1 = \Pi(1, x, y), \ldots, rx = \Pi(x, x, y)$ is a safe convention to adopt, and this is the definition which we use. (Note also that $\langle r1, \ldots, rn, 0, 0, \ldots, 0 \rangle = \langle r1, \ldots, rn, 0 \rangle$; see Exercise 2.1.5a.)

2.4.3 PROPOSITION   *Let x code a program and let i be such that $1 \leq i \leq Ln(x)$. The following functions are primitive recursive:*

1. *Nextline, where $Nextline(i, x, y)$ is the number of the next instruction executed after executing the ith instruction in the program coded by x on registers with their contents coded by y;*
2. *Nextcont, where $Nextcont(i, x, y)$ is the code of the next contents of the registers after executing the ith instruction in the program coded by x on registers with their contents coded by y;*
3. *Comp, where $Comp(x, y, m) = \langle i, z \rangle$ such that, after running the program coded by x for m steps on registers with their initial contents coded by y, i is the number of the next instruction to be executed and z codes the current contents of the registers.*

Proof

1. *Nextline(i, x, y) is $i+1$ unless* the ith instruction is a jump and the contents of the register being tested are not equal to zero. Thus

$Nextline(i, x, y) =$

$$\begin{cases} max\,j \le Ln(x)[j < i\ and\ Nam(Line(j, x)) = Jmp(Line(i, x))] \\ \quad if\quad Typ(Line(i, x)) = 4\ and\ \Pi(Reg(Line(i, x)), x, y) \ne 0 \\ min\,j \le Ln(x)[j > i\ and\ Nam(Line(j, x)) = Jmp(Line(i, x))] \\ \quad if\quad Typ(Line(i, x)) = 5\ and\ \Pi(Reg(Line(i, x)), x, y) \ne 0 \\ i + 1\ otherwise. \end{cases}$$

Notice that by this definition, if the ith instruction is the final **continue** then $Nextline$ signals that the program has halted by yielding $Nextline(i, x, y) > Ln(x)$.

2. For $Nextcont$ we need two helpful primitive recursive functions Add and Sub where $Add(j, x, y)$ codes the contents resulting from adding 1 to the contents of Rj as coded by y, and $Sub(j, x, y)$ gives the result of subtracting 1 from the contents of Rj. You can easily verify that

$$Sub(j, x, y) = min\,z \le y[\Pi(j, x, z) = \Pi(j, x, y) \doteq 1 \quad and$$
$$\forall k \le x(0 < k \ne j\ implies\ \Pi(k, x, z) = \Pi(k, x, y)];$$

we leave the definition of Add as an exercise below. Then

$Nextcont(i, x, y) =$

$$\begin{cases} Add(Reg(Line(i, x)), x, y) & if\quad Typ(Line(i, x)) = 1 \\ Sub(Reg(Line(i, x)), x, y) & if\quad Typ(Line(i, x)) = 2 \\ y & if\quad Typ(Line(i, x)) \ge 3. \end{cases}$$

3. For $Comp$ recall that Π_1 and Π_2 are the primary projection functions satisfying $\Pi_1(z) = \Pi(1, 2, z)$ and $\Pi_2(z) = \Pi(2, 2, z)$. Then $Comp$ is defined recursively as follows: $Comp(x, y, 0) = \langle 1, y \rangle$ and

$Comp(x, y, m + 1) =$

$$\langle Nextline(\Pi_1(Comp(x, y, m)), x, \Pi_2(Comp(x, y, m))),$$
$$Nextcont(\Pi_1(Comp(x, y, m)), x, \Pi_2(Comp(x, y, m)))\rangle;$$

recall that $\Pi_1(Comp(x, y, m))$ is the number of the next instruction to be executed, and that $\Pi_2(Comp(x, y, m))$ codes the current contents of the registers.

2.4.4 EXERCISE Show that the function Add in the previous proof is primitive recursive. If you get stuck you might want to use the hint in the restatement of this exercise at the end of this section; if you find no difficulty whatsoever, you might want to question whether your solution is really correct. ☐

Suppose that x codes a program. We define the *partial* function *End* so that if the program coded by x eventually halts when run with the initial contents of the registers coded by y then $End(x, y)$ is the number of steps for which this computation runs, and if the program does not halt then $End(x, y)$ is undefined. We see that *End* is a partial recursive function since

$$End(x, y) = min\ m[\Pi_1(Comp(x, y, m)) = Ln(x)].$$

Of course *End* is not primitive recursive; it is not even total.

By the results of Section 2.1 we can safely restrict our attention to the partial recursive functions of one argument, and these are computed by RAM programs with the argument as the initial contents of R1 and with the other registers initially empty. From Exercise 2.1.5a we recall that

$$\langle y, 0, 0, \ldots, 0 \rangle = \langle y, \langle 0, 0, \ldots, 0 \rangle \rangle = \langle y, 0 \rangle,$$

and so $\langle y, 0 \rangle$ is the code for having y in R1 and the rest of the registers empty. Notice that 0 and 1 do not code programs and recall that $\Pi_1(y) = \Pi(1, x, y)$ for $x > 2$, thus if x codes a program then $\Pi_1(y)$ is the contents of R1 as coded by y. Then if we define the partial recursive function ψ by

$$\psi(x, y) = \Pi_1(\Pi_2(Comp(x, \langle y, 0 \rangle, End(x, \langle y, 0 \rangle)))).$$

If x codes a program then $\psi(x, y)$ is the value of the partial recursive function of one argument computed by that program on argument y; if x does not code a program, there is nothing particularly enlightening we can say about $\psi(x, y)$. This behavior of ψ when x does not code a program is a bit troublesome, so we define the partial recursive function ϕ_{univ} by

$$\phi_{univ}(x, y) = \begin{cases} \psi(x, y) & \text{if } PROG(x) \\ undefined & \text{otherwise,} \end{cases}$$

and if x does not code a program then we know exactly what $\phi_{univ}(x, y)$ is. Notice that we have employed the notational convenience of writing ϕ_{univ} as a function of two arguments; of course, to write it as a function of one argument we would simply write $\phi_{univ}(\langle x, y \rangle)$.

ϕ_{univ} is a *universal* partial recursive function in the sense that we can obtain *any* partial recursive function of one argument from it simply by holding the first argument in ϕ_{univ} constant at an appropriate value. Also notice that the only function used in defining ϕ_{univ} which is not primitive recursive is *End*, and so we see once again (as we did in the proofs of Theorems 1.8.7 and 2.2.5) that every partial recursive function can be obtained in the form $p \circ min\ q$ where p and q are primitive recursive.

In addition, ϕ_{univ} suggests a very important point of view: using ϕ_{univ} we can interpret *every* natural number as corresponding to a RAM program, or equally well as the partial recursive function of one argument computed by that program. If x codes a program (i.e., if $PROG(x)$) and **P** is the program which x codes (i.e., '**P**' $= x$) then we call **P** the xth RAM program, and if x does not code a program then the xth RAM program is the "infinite loop" program

$$
\begin{array}{lll}
\text{N}1 & \textbf{add} & \text{R}1 \\
\text{N}1 & \text{R}1 \ \textbf{jmp} & \text{N}1\text{a} \\
\text{N}1 & \textbf{continue.} &
\end{array}
$$

We let $\mathbf{P}_x$ stand for the xth RAM program. Thus we have an *indexing*, or listing, of RAM programs $\mathbf{P}_0$, $\mathbf{P}_1$, $\mathbf{P}_2$, $\mathbf{P}_3$, ... such that every RAM program (of the restricted type we are considering in this section) with the exception of the "infinite loop" program appears in this list exactly once. Furthermore, given x we can primitive recursively compute precisely what the program $\mathbf{P}_x$ is: we first test to see if $PROG(x)$; if the answer is "no" then $\mathbf{P}_x$ is the infinite loop program; if the answer is "yes" then we use our primitive recursive functions Ln, Pg, $Line$, Typ, Nam, Reg, and Jmp to decode x. From ϕ_{univ}, or from our indexing of RAM programs, we can now define an *indexing*, or *Gödel numbering*, of the partial recursive functions of one argument: for each x, ϕ_x is the partial recursive function of one argument computed by $\mathbf{P}_x$; that is, $\phi_x(y) = \phi_{univ}(x, y)$ for all x and y. The existence of a universal function ϕ_{univ} is a fundamental property of programming systems such as RAM programs, and we honor it with the following theorem.

2.4.5 THEOREM (Enumeration, or Normal Form) *For the indexing of RAM programs defined above there is a universal partial recursive function ϕ_{univ} such that if ϕ_x is the partial recursive function of one argument computed by $\mathbf{P}_x$ then for all x and y, $\phi_x(y) = \phi_{univ}(x, y)$.*

There are some other important things to note about the universal function ϕ_{univ}. Since ϕ_{univ} is a partial recursive function, there must be a RAM program **UP** which computes *it*. Such a RAM program **UP** could be called a *universal program*, or an *interpreter* for RAM programs, since by fixing the input to register R1 at x and putting any input y in R2, **UP** will then simulate the program $\mathbf{P}_x$ on input y (note that we are viewing ϕ_{univ} as a function of two arguments here; if we were viewing it as a function of one argument we would put $\langle x, y \rangle$ in R1 to get **UP** to simulate $\mathbf{P}_x$ on input y). Thus we have made the RAM into a *stored program computer*! By giving our universal program **UP** the "program" x and the "data" y we get the result of executing program $\mathbf{P}_x$ on input data y. Moreover, since **UP** is a universal program, the halting problem for it is just as difficult as the general halting problem for all programs. In Theorem 2.4.2 we showed that there can be no algorithm for deciding whether an arbitrary program halts on an arbitrary input. The universal program **UP** is a *single* program for which no algorithm can decide whether or not it halts on an arbitrary input. This is one example of the following very general phenomenon: it is as difficult to "deal" with some single programs as it is to "deal" with the full range of arbitrary programs. In addition, from our translation results in Chapter 1, we now see that there are single universal partial recursive functions, universal Turing machines, etc., for which no algorithm can decide whether or not they halt on arbitrary inputs.

The existence of universal functions is one of the fundamental properties of acceptable programming systems. It may seem like such an obvious property that it is hardly worth mentioning, let alone being stated as a theorem. However, it is one of two fundamental properties which yield a mathematical characterization of acceptable programming systems. The other property in the characterization is the effective composition of functions, given in the next theorem.

2.4.6 THEOREM *There is a primitive recursive function c such that for all x and y, $\phi_{c(x,y)} = \phi_x \circ \phi_y$.*

Proof If x and y both code programs then $\phi_x \circ \phi_y$ can be computed by first running $\mathbf{P}_y$, clearing all registers but R1, then running $\mathbf{P}_x$. If either x or y does not code a program then $\phi_x \circ \phi_y$ is the completely undefined function, which is computed by the infinite loop program. Let n be any index for the infinite loop program (i.e., $notPROG(n)$); then a primitive recursive function c which works is defined by

$c(x, y) =$

$$\begin{cases} Conprogs(y, Conprogs(Cumclr(y), x)) & \text{if } PROG(x) \text{ and } PROG(y) \\ n & \text{otherwise.} \end{cases}$$

where *Conprogs* and *Cumclr* are from Exercise 2.4.1.

2.4.7 EXERCISE

(a) Verify in detail that the function c defined above works as claimed.

(b) Show that no matter what pairing function we use, no number less than 3 can code a program. Thus we could have taken $n = 0$ in the proof of the previous theorem. □

Additional Exercises

*2.4.4 Show that the function *Add* in the proof of Proposition 2.4.3 is primitive recursive. You may want to use the function *Large* from Exercise 2.1.5e.

2.4.8 Choose one of the specific pairing functions from Section 2.1 and find the smallest number x which codes a RAM program.

*2.4.9 Show that there is a *one-to-one* primitive recursive function p such that for all x and y, $\phi_{p(x,y)} = \phi_x$. Such a function p is often called a "padding" function.

*2.4.10 Show that there is a primitive recursive function f such that for all i and x, $\phi_{f(i)}(x)$ exists if and only if $\phi_i(x)$ exists, and if $\phi_i(x)$ exists then $\phi_{f(i)}(x)$ is the largest number stored in any of the RAM registers during the computation of program $\mathbf{P}_i$ on input x.

2.4.11 Show that there is a partial recursive function ψ which cannot be extended to a total recursive function; that is, $\psi \not\subseteq f$ for all total recursive functions f.

2.4.12 In contrast to the discussion following Theorem 1.8.7, show that there is a partial recursive function ψ which cannot be obtained directly from a total recursive function by an application of minimization; that is, there is no total recursive f such that $\psi = \min f$.

*2.4.13 Let $S_0 = \{i : \phi_i(i) = 0\}$ and $S_1 = \{i : \phi_i(i) = 1\}$. Show that there is no recursive set R such that $S_0 \subseteq R$ and $S_1 \cap R = \phi$. Sets such as S_0 and S_1 are called recursively inseparable. The recursive inseparability of S_0 and S_1 plays a key role in Chapter 4.

2.5 UNSOLVABILITY OF THE WORD PROBLEM FOR MONOIDS

In the previous section we saw that the halting problem for, say, Turing machines, is unsolvable. In addition, in the discussion following Theorem 2.4.5 on universal programs we saw that there are single Turing machines **UT** for which there is no algorithm to determine whether or not **UT** eventually halts after being given some arbitrary input string. The unsolvability of these halting problems implies the unsolvability of many other problems in mathematics and in computer science. In this (optional) section and in the next (optional) section we examine two such problems, and we prove their unsolvability by "reducing" the unsolvability of the halting problem to them. In this section it is a technical convenience to assume that we have a single Turing machine **T** which has an unsolvable halting problem as mentioned above and which has the following additional property: whenever **T** does halt the tape contains a single 0 and no other nonblank symbols.

2.5.1 EXERCISE Give a simple and uniform method for converting any Turing machine into one which halts on exactly the same inputs and whose output is 0 whenever it halts.

A *monoid* (or a semigroup with identity) is a set S together with a binary operation $\cdot : S \times S \to S$ such that

(a) $\cdot$ is *associative*, that is $(x{\cdot}y){\cdot}z = x{\cdot}(y{\cdot}z)$ for all x, y, and z in S; and
(b) there is an *identity element* 1 in S such that $x{\cdot}1 = 1{\cdot}x = x$ for all x in S.

As an example of a monoid, take S to be $A_k{}^*$ for some k, take $\cdot$ to be the operation of concatenation, and take the identity element to be ϵ, the empty word; the operation of concatenation is obviously associative.

2.5.2 EXERCISE
(a) Show that the integers under addition form a monoid, as do the integers under multiplication. In each case, what is the identity?
(b) Prove that the class of all primitive recursive functions and that the class of partial recursive functions each form a monoid with composition as "$\cdot$" and the identity function serving as the element "1".

A common method mathematicians use to get other monoids is to define an equivalence relation on $A_k{}^*$ as follows: take a fixed finite set of *relations*

$$x_1 \simeq y_1$$
$$\vdots$$
$$x_n \simeq y_n$$

where the x_i's and y_i's are in $A_k{}^*$. For w and z in $A_k{}^*$ we say that w is *directly equivalent* to z, written $w \leftrightarrow z$, if z can be obtained from w by replacing one occurrence of some x_i by the corresponding y_i or by replacing one occurrence of some y_i by the corresponding x_i. We say that w is *equivalent* to z, written $w \Leftrightarrow z$, if there are $w_0, \ldots, w_m$ in $A_k{}^*$ such that $w = w_0 \leftrightarrow \ldots \leftrightarrow w_m = z$ or if $m = 0$ and $w = z$. This relation $\Leftrightarrow$ is easily seen to be an equivalence relation on $A_k{}^*$, and its equivalence classes (which are named by the elements x in $A_k{}^*$) are given by

$$[x] = \{y : y \in A_k{}^* \quad \text{and} \quad x \Leftrightarrow y\}.$$

We can now take $S = \{[x] : x \in A_k{}^*\}$ to be the set of equivalence classes on A_k^* with respect to $\Leftrightarrow$; we define $[x]\cdot[y] = [xy]$ and we take $[\epsilon]$ to be the identity element. With these definitions, S forms a monoid.

2.5.3 EXERCISE Verify that the set of equivalence classes S with the operation and identity defined above does in fact form a monoid. Note that one thing you must check is that the operation $\cdot$ is *well defined*; that is that $[x]\cdot[y]$ does not depend on the particular "names" x and y chosen for the classes $[x]$ and $[y]$.

With the examples of monoids we have just given, it is natural to ask whether when we are given two words x and y in A_k^* we can effectively tell if they are equivalent under $\Leftrightarrow$; that is, can we tell whether x and y both name the same element $[x]$ in the monoid S? We are going to show that there is no algorithm to solve this problem. This problem is often referred to as the *word problem for monoids*. Our method is to show that the question of whether a given Turing machine **T** halts on an arbitrary input can be reduced to the question of whether certain words x and y (which are easily obtained from the input string) are equivalent under the equivalence $\Leftrightarrow$ obtained from a given set of relations $x_1 \simeq y_1, \ldots,$ $x_n \simeq y_n$. Thus any algorithm for solving the word problem for monoids could be easily converted into an algorithm for solving the halting problem for Turing machines.

We are going to show that the word problem for monoids is unsolvable by producing a *single*, fixed set of relations of the form $x_1 \simeq y_1, \ldots, x_n \simeq y_n$ such that there is no algorithm for deciding whether two words x and y are equivalent under the equivalence $\Leftrightarrow$ obtained from these relations. These relations will be obtained from a Turing machine **T** of the sort we described at the beginning of this section: the halting problem for **T** is unsolvable and **T** never gives outputs other than 0. To see where we get these relations, and why they work, we recall the proof of Theorem 1.8.3 in which we showed how Markov algorithms can simulate Turing machine computations by producing sequences of instantaneous descriptions of the computations. Suppose we take the Markov algorithm productions for the Turing machine **T** in the proof of Theorem 1.8.3 which are given *before* Exercise 1.8.4, that is those Markov productions whose correctness does not depend on the order in which they are written, and we replace the last of these productions, $r_{p+2}\# \to_t \epsilon$, by $r_{p+2}\# \to \$$. Let us call the Markov algorithm with this set of productions **M**. Then from the properties of **T** and the way **M** is constructed we have that **T** halts on an input string x in A_k^* if and only if $\mathbf{M}: \#r_0x\# \Rightarrow 0\$$. As we observed immediately following Exercise 1.8.4, the order of the productions in **M** is completely unimportant because for any of the intermediate words in this transformation, and in fact for any instantaneous description of **T**, at most one production in **M** can apply in at most one location in any such word.

Now we are not really interested in Markov algorithms here, but rather in sets of relations which define monoids. Therefore, we now need to define the set of relations we wish to consider. Let $x_1 \to y_1, \ldots, x_n \to y_n$ be the productions in **M**. Then we take $x_1 \approx y_1, \ldots, x_n \approx y_n$ as our set of relations. From the definition of the equivalence relation $\Leftrightarrow$ it is clear that if $\mathbf{M}{:}x \Rightarrow y$ then $x \Leftrightarrow y$, because each time a production in **M** is used, the relations allow us to make the same substitution. Thus if **T** halts on some input x then $\mathbf{M}{:}\#r_0x\# \Rightarrow 0\$$ and hence $\#r_0x\# \Leftrightarrow 0\$$. The converse implication also holds, but the reason why is not so clear. There are two problems. First, the relations can be applied in any order and can make substitutions anywhere in a word while Markov algorithms cannot; however, by the observation made at the end of the last paragraph, this problem does not arise in the relations derived from the productions of the Markov algorithm **M**. Second, an even greater difficulty is that relations can be applied "backwards," replacing a y_i by the corresponding x_i, which is why we did not want to include the relation $r_{p+2}\# \approx \epsilon$. Suppose that $\#r_0x\# \Leftrightarrow 0\$$ and let $w_0, \ldots, w_m$ be the *shortest* sequence of words such that

$$\#r_0x\# = w_0 \leftrightarrow \ldots \leftrightarrow w_m = 0\$.$$

Because this is the shortest such sequence, there can be no repetitions in it; if we had $w_i = w_j$ with $i < j$ then $w_0, \ldots, w_i, w_{j+1}, \ldots, w_m$ would be a shorter sequence with the same property. Now we claim that in the sequence $w_0, \ldots, w_m$ all relations are used in the "correct" order; that is, each w_{i+1} is obtained from w_i by replacing some x_j by the corresponding y_j. For the sake of a contradiction, let w_p be the last word in the sequence where this is *not* true. From the way **M** is constructed, and in particular because \$ does not occur in any x_j, p must be less than $m - 1$. Because each x_j and each y_j has *exactly one* of the symbols $r_0, \ldots, r_{p+2}$, \$ in it, each w_i has exactly one such symbol in it. Therefore if y_j is replaced by the corresponding x_j in obtaining w_{p+1} from w_p then, by the observation made at the end of the last paragraph, x_j must be replaced by y_j in obtaining w_{p+2} from w_{p+1}; that is $w_p = w_{p+2}$, which we know cannot happen. Thus all of the replacements in the sequence are made in the "correct" order. Since at most one production in **M** can apply in at most one location in any w_i, we have $\mathbf{M}{:}w_i \to w_{i+1}$ for $0 \le i < m$. Therefore $\mathbf{M}{:}\#r_0x\# \Rightarrow 0\$$ and so **T** halts on input x. Thus we have proven the converse implication and we have that **T** halts on input x if and only if

$$\#r_0x\# \Leftrightarrow 0\$.$$

Since **T** comes from a "universal" Turing machine, there is no algorithm to decide whether $\#r_0x\# \Leftrightarrow 0\$$ under our chosen set of relations. We have proven the following theorem.

2.5.4 THEOREM (Unsolvability of the Word Problem for Monoids) *There is no algorithm which, when given a finite set of relations and two words, can decide whether the words are equivalent under the relations. In fact, there is a single, fixed set of relations and a single fixed word $0\$$ such that there is no algorithm for deciding whether or not any given word is equivalent to $0\$$ under the relations.*

In summary, the basic outline of the proof of this theorem goes like this: if there were an algorithm for deciding whether arbitrary words are equivalent to $0\$$ under our set of relations then we could convert that into an algorithm for deciding the halting problem for **T**. The algorithm for the halting problem for **T** would simply take any input string x and convert it to the string $\#r_0x\#$ and then ask the algorithm for the word problem whether $\#r_0x\#$ is equivalent to $0\$$. Since we know there is no algorithm for the halting problem for **T**, there can be no algorithm for solving the word problem for monoids.

Additional Exercises

2.5.5 Show that the word problem for monoids over the alphabet $A_1 = \{a_1\}$ is *solvable*; that is, give an algorithm which when given a finite set of relations and two words will decide whether or not the two words are equivalent under the relations.

2.5.6 Using appropriate codings, show that the word problem for monoids over any alphabet A_k with $k \geq 2$ is unsolvable.

2.5.7 Formulate the word problem for *groups*, and prove that it is unsolvable. (*Caution*: this is generally considered to be a hell of a lot longer and harder than the proof for monoids.)

2.6   THE UNSOLVABILITY
OF THE POST CORRESPONDENCE PROBLEM

We saw in the previous section that one way to show that problems are algorithmically unsolvable is to reduce the halting problem to them. Another fundamental unsolvable problem is the Post Correspondence Problem, and it too is used very frequently for showing problems to be algorithmically unsolvable by reducing it to a problem one wishes to show unsolvable. It is particularly helpful in formal language theory. In this (optional) section we prove the unsolvability of the Post Correspondence Problem. Of course, we do this by reducing the halting problem (specifically the halting problem for Turing machines) to the Post Correspondence Problem.

A *Post Correspondence System* (*over A_k^**) is a finite sequence

$$x_1 \leftrightarrow y_1$$
$$\vdots$$
$$x_n \leftrightarrow y_n$$

of *correspondences*, where $x_1, \ldots, x_n, y_1, \ldots, y_n$ are in A_k^*. The *Post Correspondence Problem* is the problem of deciding, for any Post Correspondence System, whether or not there is a *solution* consisting of a finite sequence of integers $i_1, \ldots, i_m$ such that

$$x_{i_1} x_{i_2} \ldots x_{i_m} = y_{i_1} y_{i_2} \ldots y_{i_m}.$$

For example, consider the correspondence system

$$0011 = x_1 \leftrightarrow y_1 = 0$$
$$0011 = x_2 \leftrightarrow y_2 = 0110$$
$$1 = x_3 \leftrightarrow y_3 = 0110.$$

It is obvious to the most casual observer that no solution can begin with x_2 and y_2 or with x_3 and y_3: therefore any solution must begin with x_1 and y_1. If after beginning with x_1 and y_1 we continue the sequence with x_1 and y_1 or with x_3 and y_3, then the sequence could not possibly be extended to a solution (why?); so we must use x_2 and y_2 next, yielding

x-list	0011	0011
y-list	0	0110.

For the same reasons as above, we cannot continue these lists with x_1 and y_1 or with x_3 and y_3: so we must use x_2 and y_2 again, yielding

x-list	0011	0011	0011
y-list	0	0110	0110.

We now see that we are in an "infinite loop" with a repeating pattern that will never terminate properly: the y-list is always "chasing" the x-list. Therefore, this particular Post Correspondence System has no solution.

Before proving that there is no algorithm for solving the Post Correspondence Problem, we give a simple application of the unsolvability of this problem. We assume the unsolvability of the Post Correspondence Problem and show that there can be no algorithm to decide whether the intersection of two (linear) context-free languages is empty. We do this by showing that if there were such an algorithm then we could use it to solve the Post Correspondence Problem. Let $x_1 \leftrightarrow y_1, \ldots, x_n \leftrightarrow y_n$ be any Post Correspondence System. Let G_x be the grammar with productions $S \rightarrow 1x_1, \ldots, S \rightarrow nx_n, S \rightarrow 1Sx_1, \ldots, S \rightarrow nSx_n$, and let G_y be the grammar with productions $S \rightarrow 1y_1, \ldots, S \rightarrow ny_n$, $S \rightarrow 1Sy_1, \ldots, S \rightarrow nSy_n$. (We are assuming that $N \cap A_k = \emptyset$.) Then $L(G_x)$, the language generated by G_x, is clearly the set of all possible strings $i_m \ldots i_1 x_{i_1} \ldots x_{i_m}$ and $L(G_y)$ is the set of all possible strings $i_m \ldots i_1 y_{i_1} \ldots y_{i_m}$, and so $L(G_x) \cap L(G_y) \neq \phi$ if and only if the correspondence system has a solution. Therefore any algorithm which determined whether or not the intersection of two (linear) context-free languages was empty could be used to solve the Post Correspondence Problem.

We now show that the Post Correspondence Problem is itself unsolvable. For each Turing machine and input string we give a correspondence system which "simulates" the Turing machine in such a way that the system will have a solution if and only if the Turing machine halts on the given input string. Thus any algorithm to solve the correspondence problem would also solve the halting problem for Turing machines.

Let **T** be a Turing machine with alphabet A_k and states $0, \ldots, p$. Recall how in our construction of Markov algorithms to simulate Turing machines in the proof of Theorem 1.8.3 we represented the situation of **T** being in state i looking at a_j on the tape "$\ldots a_j \ldots$" by the instantaneous description (*ID*) "$\ldots r_i a_j \ldots$". Our correspondence system will have only one possible sequence that could lead to a solution (just as in our example), and this sequence will be the series of instantaneous descriptions of the computation of **T** on the given input. We arrange things (similarly to the example) so that the x-list will always have one fewer instantaneous description on it than the y-list, and as we add to the x-list to match the last instantaneous description on the y-list we must build the next *ID* on the y-list. Then we add correspondences so that if **T**'s computation ever halts we are able to finish the two lists off to be the same, a solution! Note that if any correspondence has $x_i = y_i$ then the system has a trivial solution given by the sequence i yielding the string x_i. To avoid this trivial solution, we use both barred and unbarred characters, and alternate instantaneous descriptions in the lists will be barred. Finally, let $\#$, q, and r be new symbols; $\#$ will be used as a marker to separate instantaneous descriptions.

Our system will have the following correspondences:

$$a_1 \leftrightarrow \bar{a}_1 \qquad \bar{a}_1 \leftrightarrow a_1 \qquad \# \leftrightarrow \bar{\#}$$
$$\vdots \qquad\qquad \vdots \qquad\qquad \bar{\#} \leftrightarrow \#$$
$$a_k \leftrightarrow \bar{a}_k \qquad \bar{a}_k \leftrightarrow a_k$$

and for each instruction $i \, a_j \, a_n \, R \, m$ the correspondences

$$r_i a_j a_1 \leftrightarrow \bar{a}_n \bar{r}_m \bar{a}_1 \qquad \bar{r}_i \bar{a}_j \bar{a}_1 \leftrightarrow a_n r_m a_1 \qquad r_i a_j \bar{\#} \leftrightarrow \bar{a}_n \bar{r}_m \bar{a}_k \#$$
$$\vdots \qquad\qquad \vdots \qquad\qquad \bar{r}_i \bar{a}_j \# \leftrightarrow a_n r_m a_k \#$$
$$r_i a_j a_k \leftrightarrow \bar{a}_n \bar{r}_m \bar{a}_k \qquad \bar{r}_i \bar{a}_j \bar{a}_k \leftrightarrow a_n r_m a_k$$

(recall that $a_k = B$ is a blank); and for each instruction $i \, a_j \, a_n \, L \, m$ the correspondences

$$a_1 r_i a_j \leftrightarrow \bar{r}_m \bar{a}_1 \bar{a}_n \qquad \bar{a}_1 \bar{r}_i \bar{a}_j \leftrightarrow r_m a_1 a_n \qquad \# r_i a_j \leftrightarrow \# \bar{r}_m \bar{a}_k \bar{a}_n$$
$$\vdots \qquad\qquad \vdots \qquad\qquad \# \bar{r}_i \bar{a}_j \leftrightarrow \# r_m a_k a_n$$
$$a_k r_i a_j \leftrightarrow \bar{r}_m \bar{a}_k \bar{a}_n \qquad \bar{a}_k \bar{r}_i \bar{a}_j \leftrightarrow r_m a_k a_n$$

and to start the lists correctly for input x, the correspondence

$$\# \leftrightarrow \# a_k r_0 x \bar{\#}$$

These correspondences "simulate" the computation of **T** on input x. Notice that the correspondence $\# \leftrightarrow \# a_k r_0 x \bar{\#}$ gives the *only* possible way to start a solution, so that in the beginning the x-list will have only an end marker, $\#$, while the y-list will have a complete instantaneous description. In general this pattern will prevail, with the x-list always "chasing" the y-list. In the process, while the x-list is "copying" the last instantaneous description from the y-list, the y-list must be building the next instantaneous description. Now to finish off the lists in case **T** halts, for each pair $i \, a_j$ such that **T** has no instruction beginning $i \, a_j \ldots$ we add the correspondences

$$r_i a_j \leftrightarrow \bar{q} \qquad \bar{r}_i \bar{a}_j \leftrightarrow q$$

and also the correspondences

$$q a_1 \leftrightarrow \bar{q} \qquad \bar{q} \bar{a}_1 \leftrightarrow q \qquad q \bar{\#} \leftrightarrow \bar{r} \# \qquad a_1 r \leftrightarrow \bar{r} \qquad \bar{a}_1 \bar{r} \leftrightarrow r$$
$$\vdots \qquad \vdots \qquad q \# \leftrightarrow r \# \qquad \vdots \qquad \vdots$$
$$q a_k \leftrightarrow \bar{q} \qquad \bar{q} \bar{a}_k \leftrightarrow q \qquad\qquad a_k r \leftrightarrow \bar{r} \qquad \bar{a}_k \bar{r} \leftrightarrow r$$

If and when **T** halts, these correspondences "erase" **T**'s output and give us a y-list ending with either "$\ldots \# r \bar{\#}$" or "$\ldots \bar{\#} \bar{r} \#$", while the x-list goes only as far as "$\ldots \#$" or "$\ldots \bar{\#}$", respectively. Thus to finish the lists off equally and have a solution we need the correspondences

$$r \bar{\#} \leftrightarrow \epsilon \qquad \bar{r} \# \leftrightarrow \epsilon.$$

2.6.1 EXERCISE Verify that the Post Correspondence system we have given has a solution if and only if the Turing machine **T** halts on input x. (The key step is to prove by induction on n that as long as **T** has not halted, the only possible x-list and y-list that could lead to a solution looks like

$$x\text{-list} \quad \# ID_0 \# \overline{ID_1} \# \ldots \# ID_{2n} \#$$

$$y\text{-list} \quad \# ID_0 \# \overline{ID_1} \# \ldots \# ID_{2n} \# \overline{ID_{2n+1}} \#$$

where ID_j is an instantaneous description of $\mathbf{T}$ after j steps in its computation; for example $ID_0 = a_k r_0 x$.)

We have proved the final theorem of this chapter.

2.6.2 THEOREM   *There is no algorithm for solving the Post Correspondence Problem.*

Additional Exercises

2.6.3   Suppose we allow infinite solutions to Post Correspondence Systems; that is, suppose that we allow a solution to be given either by a finite sequence of integers as before or by an infinite sequence of integers $i_1, i_2, \ldots, i_n, \ldots$ such that

$$x_{i_1} x_{i_2} \ldots x_{i_n} \ldots = y_{i_1} y_{i_2} \ldots y_{i_n} \ldots .$$

Is this extended Post Correspondence Problem solvable? Prove your answer.

2.6.4   Let $x_1 = ab$, $x_2 = baa$, $x_3 = aba$, $y_1 = aba$, $y_2 = aa$, and $y_3 = baa$.

(a) Is $\{x_1, x_2, x_3\}^* \cap \{y_1, y_2, y_3\}^*$ empty?
(b) Does the correspondence system $x_1 \leftrightarrow y_1$, $x_2 \leftrightarrow y_2$, $x_3 \leftrightarrow y_3$ have a solution? Prove your answer.

2.6.5   Give an algorithm which, when given two lists of words $x_1, \ldots, x_m$ and $y_1, \ldots, y_n$ over the same alphabet, decides whether or not $\{x_1, \ldots, x_m\}^* \cap \{y_1, \ldots, y_n\}^*$ is empty. Note that this amounts to deciding whether there are *two* sequences of integers $i_1, \ldots, i_p$ and $j_1, \ldots, j_q$ such that

$$x_{i_1} \ldots x_{i_p} = y_{j_1} \ldots y_{j_q}.$$

Chapter 3
Basic Recursive Function Theory

In this chapter we present some very basic theory of the partial recursive functions. Although mathematicians have done a great deal of interesting and important work on the theory of the partial recursive functions, in this chapter we restrict ourselves to developing just the bare minimum that is necessary for the theory of algorithms. The main results of this chapter all play an important role in our understanding of the theory of algorithms, and many of them have important applications in the further development of the theory. In Section 3.1 we present the very general notion of an acceptable programming system. We justify studying this notion by showing that there are effective translations between any two such systems, and hence there are effective translations between any such system and any "reasonable" programming system, such as those we studied in Chapter 1. In Section 3.2 we demonstrate the unsolvability of some fundamental problems concerning acceptable programming systems. We begin with the halting problem and conclude with the "all encompassing" Rice's Theorem, which asserts that there are no algorithms for testing nontrivial properties of the input-output behavior of programs. In Section 3.3 we characterize the intuitive notion of "partially solvable" problems with the technical notion of recursively enumerable sets, that is, sets for which there are algorithms which *list* their members. We then give an Extended Rice's Theorem which characterizes those properties of the input-output behavior of programs which are "partially solvable." In the last section of this chapter we prove two very fundamental theorems about acceptable programming systems. The Recursion Theorem justifies the use of extremely general types of recursive definitions of functions in any acceptable programming system, without having to have such facilities explicitly built in. This theorem is an important tool for developing the theory of algorithms. We conclude the chapter with Rogers' Isomorphism Theorem, which shows that between any two acceptable programming systems there is an effective, one-to-one and onto translation, or "isomorphism"; hence there is a very strong sense in which all acceptable programming systems are "equivalent."

3.1 ACCEPTABLE PROGRAMMING SYSTEMS

Throughout this chapter, and throughout most of the rest of this book, we shall rely on the results of Sections 2.1 and 2.2, restricting our attention to the partial recursive functions of one argument over the natural numbers. We remind the reader of our assertion at the end of Section 2.2 that while this restriction provides for considerable simplification of *notation* and *terminology* in the development of our theory, there are no *conceptual* difficulties in developing the theory for all partial recursive functions over arbitrary alphabets directly, rather than via coding and decoding. Recall that functions of several arguments are "coded" into functions of a single argument by composing them with appropriate projection functions, and so although we are *formally* restricting ourselves to functions of a single argument, through the use of simple projection functions we have the ability to work *intuitively* with functions of several arguments. For example, if we wish to define some function f which we are thinking of as having two arguments, we can simply define $f(\langle x,y \rangle)$ to have the value we would like to assign to $f(x,y)$. For the purposes of computing f, we first apply simple projection functions to an input argument $\langle x,y \rangle$ to recover x and y, then proceed as if we had been given x and y as separate input arguments. Having now carefully explained this situation, we shall often find it notationally convenient simply to write $f(x,y)$ instead of the more cumbersome $f(\langle x,y \rangle)$, and we are sure that this will not result in any confusion on the part of the reader.

Recall from Theorem 2.4.5 that for the RAM programming system the programs can be numbered in such a way that if ϕ_i is the partial recursive function computed by the ith program then the function ϕ_{univ} such that $\phi_{univ}(i,x) = \phi_i(x)$ for all i and x is a partial recursive function; that is, ϕ_{univ} is a universal partial recursive function for the RAM programming system. Furthermore, Theorem 2.4.6 showed that composition in the RAM programming system is effective since there is a total recursive function c such that $\phi_{c(i,j)} = \phi_i \circ \phi_j$ for all i and j.

Suppose that we take any of the other programming systems discussed in Chapter 1 (Turing machines, Markov algorithms, or definitions of partial recursive functions), then we could similarly number the programs in any one of these systems to produce another indexing ψ_0, $\psi_1, \ldots$ of the partial recursive functions. Since RAM programs can be translated effectively into any one of these systems, there is some total recursive function f which gives such a translation; that is $\psi_{f(i)} = \phi_i$ for all i. Similarly, since the system $\psi_0, \psi_1, \ldots$ can be translated effectively into the RAM system $\phi_0, \phi_1, \ldots$, there is a total recursive function g such that $\phi_{g(i)} = \psi_i$ for all i. But then the partial function ψ_{univ} such that

$$\psi_{univ}(i, x) = \psi_i(x) = \phi_{g(i)}(x) = \phi_{univ}(g(i), x)$$

is a universal partial recursive function for the system $\psi_0, \psi_1, \ldots$, and the total recursive function d defined by $d(i, j) = f(c(g(i), g(j)))$ is an effective function for composition in the system $\psi_0, \psi_1, \ldots$ since

$$\psi_{d(i,j)} = \phi_{c(g(i),g(j))} = \phi_{g(i)} \circ \phi_{g(j)} = \psi_i \circ \psi_j .$$

Thus indexings arising from any of the programming systems in Chapter 1 will have universal partial recursive functions and functions for effective composition, and this conclusion follows from the existence of effective translations between those systems and the RAM system.

It is probably impossible to imagine a "reasonable" programming system for which there is no effective interpretation of the programs by translation into the RAM system, or into which RAM programs cannot be effectively translated. And so by the argument we gave in the previous paragraph, *any* "reasonable" programming system must have a universal partial recursive function and a total recursive function for composition. Another way to see that this must be so is to imagine carrying out a coding of programs for any other programming system similar to the coding in Section 2.4, and then proving Theorems 2.4.5 and 2.4.6 for the indexing that is produced. So, surely every "reasonable" programming system has a universal effective method for interpreting programs and an effective method for "composing" programs. What is remarkable is that these two simple properties of all "reasonable" programming systems, having a universal partial recursive function and having a total recursive function for composition, provide a simple, useful, and elegant *characterization* of acceptable programming systems. By developing a theory of such systems we have results which apply to *all* "reasonable" programming systems, including all of the programming languages currently in use or being proposed. The next definition gives this notion of acceptable programming systems.

3.1.1 DEFINITION A *programming system* is a listing $\phi_0, \phi_1, \ldots$ which includes *all* of the partial recursive functions (of one argument over N). A programming system $\phi_0, \phi_1, \ldots$ is *universal* if the partial function ϕ_{univ} such that $\phi_{univ}(i, x) = \phi_i(x)$ for all i and x is itself a partial recursive function; that is, if the system has a universal partial recursive function. A universal programming system $\phi_0, \phi_1, \ldots$ is *acceptable* if there is a total recursive function c for composition such that $\phi_{c(i,j)} = \phi_i \circ \phi_j$ for all i and j.

Programming systems are often referred to elsewhere in the literature as *indexings* or *Gödel numberings* of the partial recursive functions. From Section 2.4, and in particular from Theorems 2.4.5 and 2.4.6, we know that there is at least one acceptable programming system, the one we obtained from RAM programs in that section. Prior to Definition

3.1.1 we argued that any "reasonable" programming system satisfies our definition of an acceptable programming system. Therefore, any results we prove about acceptable programming systems should hold for *all* "reasonable" programming systems, and certainly for all existing general purpose programming languages.

An important and useful property of reasonable programming systems is the ability to modify programs so that some input parameters are held constant, or in other words, to call parameters by value as well as by name. The next theorem, usually honored with the elegant name "the *s-m-n* Theorem," shows that this property holds for all acceptable programming systems.

3.1.2 *s-m-n* THEOREM *For any acceptable programming system ϕ_0, $\phi_1, \ldots$ there is a total recursive function s such that for all i, all $m \geq 1$ and $n \geq 1$, and for all $x_1, \ldots, x_m$ and $y_1, \ldots, y_n$,*

$$\phi_{s(i,m,x_1,\ldots,x_m)}(y_1, \ldots, y_n) = \phi_i(x_1, \ldots, x_m, y_1, \ldots, y_n).$$

That is, the function s allows us to specify that the first m arguments for the ith "program" be held constant at $x_1, \ldots, x_m$.

Proof First notice that by our coding of functions of several arguments into functions of one argument, another notation for the property of the function *s* reads

$$\phi_{s(i,m,\langle x_1,\ldots,x_m\rangle)}(\langle y_1, \ldots, y_n \rangle) = \phi_i(\langle x_1, \ldots, x_m, y_1, \ldots, y_n \rangle).$$

3.1.3 EXERCISE Using the definition of our functions $\langle , \ldots, \rangle_n$, explain why the function *s* does not need to have *n* as an argument.

Recalling the primitive recursive function *Con* from Exercise 2.1.7 which has the property that

$$Con(m, \langle x_1, \ldots, x_m \rangle, \langle y_1, \ldots, y_n \rangle) = \langle x_1, \ldots, x_m, y_1, \ldots, y_n \rangle,$$

we see that a total recursive function *s* such that for all *i*, all $m \geq 1$, and all *x* and *y*,

$$\phi_{s(i,m,x)}(y) = \phi_i(Con(m, x, y))$$

will satisfy the conditions of the theorem.

To finish the proof of the *s-m-n* Theorem, we define three helpful total recursive functions *P*, *Q*, and *R*. We let $P(y) = \langle 0, y \rangle$, and $Q(\langle x, y \rangle) = \langle x+1, y \rangle$ for all *x* and *y*. Since we have a programming system, there are "programs" *p* and *q* such that $\phi_p = P$ and $\phi_q = Q$. We define $R(0) = p$ and $R(x+1) = c(q, R(x))$ for all *x*, where *c* is a total recursive function for composition in our acceptable programming system.

3.1.4 EXERCISE Show that $\phi_{R(x)}(y) = \langle x, y \rangle$ for all x and y. If you have a little trouble with this exercise, consult its restatement at the end of this section.

Recalling that by definition, $\langle x, y, z \rangle = \langle x, \langle y, z \rangle \rangle$, we notice that

$$\phi_{R(x)} \circ \phi_{R(y)}(z) = \phi_{R(x)}(\langle y, z \rangle) = \langle x, y, z \rangle.$$

Finally, let k be such that $\phi_k(\langle m, x, y \rangle) = Con(m, x, y)$, and define

$$s(i, m, x) = c(i, c(k, c(R(m), R(x)))).$$

Then we have

$$\phi_{s(i,m,x)}(y) = \phi_i \circ \phi_k \circ \phi_{R(m)} \circ \phi_{R(x)}(y) = \phi_i(Con(m, x, y)),$$

which completes the proof of the s-m-n Theorem. Notice that if the function c is primitive recursive then s will also be primitive recursive, and at least intuitively, s will not be much harder to compute than c. $\square$

The preceding proof shows how to construct s-m-n functions using a universal function and a function for composition. In *reasonable* programming systems, s-m-n functions can be constructed *directly* by describing the very simple manipulations on programs needed to obtain the desired results. See the construction of the composition function for our RAM programming system in the proof of Theorem 2.4.6. After all, s-m-n functions simply supply some constants as some of the inputs to programs.

Our definition of acceptable programming systems is not generally used in the literature; alternative definitions are far more common. One such definition defines an acceptable programming system to be a universal programming system into which there is an effective translation of some fixed programming system such as our RAM system. Another definition defines an acceptable programming system to be a universal programming system which satisfies the s-m-n Theorem. These two alternative definitions are the ones most commonly used in the literature, and by working Exercises 3.1.7 and 3.1.8 at the end of this section you will see that they are each equivalent to our definition.

We saw in Chapter 1 that for any two of the programming systems considered there, either can be effectively translated into the other. This is a very general phenomenon. Indeed it follows from our next theorem that for *any* two acceptable programming systems, each can be effectively translated into the other.

3.1.5 THEOREM Let $\phi_0, \phi_1, \ldots$ be any universal programming system, and let $\psi_0, \psi_1, \ldots$ be any programming system with a total *recursive s-1-1 function: that is, there is a total recursive function s such that for all i, x, and y*

$$\psi_{s(i,x)}(y) = \psi_i(x, y).$$

Then there is a total recursive function t which translates the system ϕ_0, ϕ_1, . . . into the system ψ_0, ψ_1, . . . ; that is, $\phi_i = \psi_{t(i)}$ for all i.

Proof Let ϕ_{univ} be a universal partial recursive function for the system ϕ_0, ϕ_1, ... Since the programming system ψ_0, ψ_1, . . . contains all partial recursive functions, there is a fixed "program" k such that $\phi_{univ} = \psi_k$. Then

$$\psi_{s(k,i)}(x) = \psi_k(i, x) = \phi_{univ}(i, x) = \phi_i(x)$$

for all i and x. Therefore, the desired translation function t can simply be defined by $t(i) = s(k, i)$ for all i.   □

If we have any two acceptable programming systems, then the previous theorem applies to them, in either order; an s-1-1 function is gotten by simply holding the second argument of an s-m-n function constant at 1. Thus any two acceptable programming systems are equivalent in a fairly strong sense: we can effectively translate either one into the other. It is worth noting about the proof of the previous theorem and the proof of the s-m-n Theorem, that if the composition function c for an acceptable programming system is primitive recursive, then the s-m-n function will also be primitive recursive, and hence every universal programming system can be translated into such an acceptable programming system by some primitive recursive function.

The preceding theorem also provides a helpful technical tool for proving properties of acceptable programming systems; it is often convenient to prove a property for a particular acceptable programming system and then "transfer" the property to any other acceptable programming system by using translating functions. For example,

3.1.6 EXERCISE Let ψ_0, ψ_1, . . . be any acceptable programming system. Show that there is a total recursive function *step* such that for all i and x:

(a) There is an m such that $step(i, x, m) \neq 0$ if and only if $\psi_i(x)$ exists.

(b) If $step(i, x, m) \neq 0$, then $step(i, x, m) = \psi_i(x) + 1$.

Functions such as *step* are sometimes called "step counting" functions.

Additional Exercises

*3.1.4 Prove by induction on x that $\phi_{R(x)}(y) = \langle x,y \rangle$ where R is the function from the proof of the s-m-n Theorem.

3.1.7 Let ϕ_0, ϕ_1, . . . be our RAM programming system from Section 2.4, and let ψ_0, ψ_1, . . . be any universal programming system. Show that if there is a total recursive function f translating the RAM system into ψ_0, ψ_1, . . . (that is, $\phi_i = \psi_{f(i)}$ for all i), then ψ_0, ψ_1, . . . is actually an acceptable programming system.

3.1.8 Show that if ψ_0, ψ_1, . . . is a universal programming system with a total recursive s-1-1 function s (that is, $\psi_{s(i,x)}(y) = \psi_i(x, y)$), then ψ_0, ψ_1, . . . is actually an acceptable programming system.

*__3.1.9__ Reprove Exercise 2.4.13 for arbitrary programming systems. That is, prove that if ϕ_0, ϕ_1, . . . is a listing of all partial recursive functions (not necessarily even universal), then the sets $S_0 = \{i : \phi_i(i) = 0\}$ and $S_1 = \{i : \phi_i(i) = 1\}$ are recursively inseparable.

3.2 ALGORITHMICALLY UNSOLVABLE PROBLEMS

From now on, unless specifically stated otherwise, we assume that ϕ_0, ϕ_1, . . . is any acceptable programming system. If ψ is a partial function we say that $\psi(x)$ is *convergent* if $\psi(x)$ exists (i.e., if x is in the domain of ψ) and we say that $\psi(x)$ is *divergent* if $\psi(x)$ does not exist. Recall that this terminology comes from thinking computationally, and waiting for the computation of ψ on input x to "converge," that is, halt.

In this section we show that certain general and fundamental problems concerning acceptable programming systems are algorithmically unsolvable. We begin with the halting problem. In Section 2.4 we saw that the halting problem for RAM programs is algorithmically unsolvable; there is no algorithm for deciding whether or not a given program halts on a given input. In our next theorem we show that this problem is in fact algorithmically unsolvable for *every* programming system.

3.2.1 THEOREM (Unsolvability of the Halting Problem) *Let ψ_0, ψ_1, . . . be any programming system. The function f such that for all x and y*

$$f(x, y) = \begin{cases} 1 & \text{if} \quad \psi_x(y) \text{ is convergent} \\ 0 & \text{if} \quad \psi_x(y) \text{ is divergent} \end{cases}$$

is not a recursive function.

Proof Define the total function g by $g(x) = f(x,x)$, and define the partial function θ by

$$\theta(x) = \begin{cases} 0 & \text{if} \quad g(x) = 0 \\ \text{divergent} & \text{if} \quad g(x) = 1. \end{cases}$$

We claim that θ cannot be a partial recursive function. If it were, then there would be a program i such that $\psi_i = \theta$, and then we would have that $\psi_i(i) = \theta(i) = 0$ if and only if $g(i) = f(i, i) = 0$ if and only if $\psi_i(i)$ is divergent, by the definitions of f, g, and θ. This contradiction shows that θ is not a partial recursive function, and hence f and g cannot be recursive functions either. Notice that in this proof we have not needed to assume that ψ_0, ψ_1, . . . is an acceptable, or even a universal, programming system. $\square$

The function g in the previous proof is the characteristic function of a set. We use K to denote this set; that is,

$$K = \{x : \phi_x(x) \text{ is convergent}\}$$

for any given (acceptable) programming system ϕ_0, ϕ_1, ... This "diagonal" set K is an abbreviated, or restricted, form of the halting problem, and the proof of the previous theorem shows that K is not recursive. (Recall that a set is a unary predicate, and that a predicate is recursive just in case its characteristic function is. Thus a recursive set is just a set whose characteristic function is recursive.) The set K is so important and useful that from now on, when we refer to the halting problem, unless stated otherwise, we shall be referring to the particular set K. We make frequent use of K in showing that sets and functions are not recursive by "reducing" K to them in ways similar to the use of the function f in the proof of the previous theorem. In fact, we use the unsolvability of K in the proof of the next proposition to show that certain natural questions about the input-output behavior of programs are algorithmically unsolvable.

3.2.2 PROPOSITION *For all y and z, the following sets are not recursive:*

$$A = \{x : \phi_x \text{ is a constant function}\};$$
$$B(y) = \{x : y \text{ is in the range of } \phi_x\};$$
$$C(y, z) = \{x : \phi_x(y) = z\}.$$

Note that A represents the set of programs with constant output, $B(y)$ represents the set of programs which on some input give the particular output y, and $C(y,z)$ represents the set of programs which give the particular output z on the particular input y. The proposition asserts that none of these properties of program behavior can be tested algorithmically.

Proof The proof is by "reducing" K to each of these sets. We first note that there is a total recursive function f such that

$$\phi_{f(x,y)}(z) = \begin{cases} y & \text{if } \phi_x(x) \text{ is convergent} \quad (\text{i.e., if } x \in K) \\ \text{divergent} & \text{if } \phi_x(x) \text{ is divergent} \quad (\text{i.e., if } x \notin K) \end{cases}$$

for all x, y, and z. We construct such an f by a simple application of the *s-m-n* Theorem similar to one used in the proof of Theorem 3.1.5. This type of simple *s-m-n* constructions is a *fundamental tool* we shall use many times in the remainder of this book, and the reader is cautioned to begin mastering it now. In order to obtain f, we define the partial function θ by $\theta(x, y, z) = [\phi_{univ}(x, x) \doteq \phi_{univ}(x, x)] + y$ where ϕ_{univ} is a universal partial recursive function for our programming system. From this definition of θ we have that

$$\theta(x, y, z) = \begin{cases} y & \text{if } x \in K \\ \text{divergent} & \text{if } x \notin K. \end{cases}$$

Let i be a program such that $\phi_i = \theta$, and define $f(x, y) = s(i, 2, \langle x, y \rangle)$ where s is a total recursive *s-m-n* function for our programming system. Then we have that for all x, y, and z

$$\phi_{f(x,y)}(z) = \phi_{s(i,2,\langle x,y \rangle)}(z) = \phi_i(x, y, z) = \theta(x, y, z),$$

which is precisely what we wanted. Now note that for all y we have that if x is in K then $\phi_{f(x,y)}$ is the *constant* function y, and that if x is not in K then $\phi_{f(x,y)}$ is the *totally undefined* function, $\emptyset$.

Given the function f, the remainder of the proof is quite easy. For all x, we have that x is in K if and only if $f(x,0)$ is in A; or in other words (and symbols), $c_K(x) = c_A(f(x,0))$. Intuitively, a program for deciding membership in A could easily be converted into one for deciding membership in K; to decide whether x is in K first compute $f(x,0)$ and then ask whether $f(x,0)$ is in A. Thus if A were recursive, K would also have to be recursive (c_A recursive implies c_K recursive), which we know is not the case. Similarly, we have that x is in K if and only if $f(x,y)$ is in $B(y)$ if and only if $f(x,z)$ is in $C(y,z)$; in other words, $c_K(x) = c_{B(y)}(f(x,y)) = c_{C(y,z)}(f(x,z))$. Thus $B(y)$ and $C(y, z)$ cannot be recursive either. ☐

The next theorem generalizes the previous proposition by showing that in any acceptable programming system there are *no* nontrivial properties of the input-output behavior of programs, that is, properties of partial recursive *functions*, which can be decided by looking at the *programs*. If $\mathcal{C}$ is any class of partial recursive *functions*, we define the set of *programs* $P_\mathcal{C}$ by

$$P_\mathcal{C} = \{x : \phi_x \in \mathcal{C}\}.$$

Thus if $\mathcal{C}$ is any "property" of the input-output behavior of programs, $P_\mathcal{C}$ is the set of all programs with that type of behavior. The statement of the previous proposition together with the explanation immediately following it give three examples of sets $P_\mathcal{C}$ together with their corresponding classes $\mathcal{C}$.

3.2.3 RICE'S THEOREM $P_{\mathscr{C}}$ *is recursive if and only if* $P_{\mathscr{C}} = \emptyset$ *or* $P_{\mathscr{C}} = N.$

Proof N and $\emptyset$ are certainly recursive sets. For the converse, suppose that $\emptyset \neq P_{\mathscr{C}} \neq N$. Then $\mathscr{C}$ is neither empty nor the class of all partial recursive functions. Since the recursive sets are closed under complementation, without loss of generality we may assume that $\emptyset$ (the totally undefined *function*) is *not* in $\mathscr{C}$, and that ψ is some *other* partial recursive function which *is* in $\mathscr{C}$. Again, using a standard application of the *s-m-n* Theorem we can produce a total recursive function f such that for all x

$$\phi_{f(x)} = \begin{cases} \psi & \text{if } x \in K \\ \emptyset & \text{if } x \notin K. \end{cases}$$

To obtain f let the program i compute ψ, that is $\psi = \phi_i$, and define the partial recursive function θ by

$$\theta(x, y) = \phi_{univ}(i, y) + [\phi_{univ}(x, x) \dot- \phi_{univ}(x, x)]$$

where ϕ_{univ} is a universal partial recursive function. Then if j is such that $\theta = \phi_j$ and we define $f(x) = s(j, 1, x)$ with s an *s-m-n* function, we have

$$\phi_{f(x)}(y) = \phi_{s(j,1,x)}(y) = \phi_j(x, y) = \theta(x, y) = \begin{cases} \psi(y) & \text{if } x \in 2 \\ \text{divergent} & \text{if } x \notin K. \end{cases}$$

Having the required f, the remainder of the proof is quite easy. For each x, x is in K if and only if $f(x)$ is in $P_{\mathscr{C}}$; that is, $c_K = c_{P_{\mathscr{C}}} \circ f$. Since f is recursive and c_K is *not* recursive, $c_{P_{\mathscr{C}}}$ cannot be recursive. Intuitively, a program for deciding membership in $P_{\mathscr{C}}$ could easily be converted into one for deciding membership in K; to decide whether x is in K first compute $f(x)$ and then ask whether $f(x)$ is in $P_{\mathscr{C}}$. Therefore $P_{\mathscr{C}}$ cannot be recursive since K is not.   □

Rice's Theorem is very important since it dashes all hopes of algorithmically testing input-output behavior of arbitrary programs. By doing so, it begins to delimit the contexts in which some form of such testing might be possible. Conceiving and developing approaches to programming which overcome some of the limitations expressed by Rice's Theorem is one of the goals of computer science.

Additional Exercises

3.2.4 Without using Rice's Theorem, show that there are no algorithms for testing the following properties of programs:
(a) whether a program computes the totally undefined function
(b) whether the function a program computes has an infinite domain

(c) whether the function a program computes is total
(d) whether two programs compute the same function

3.2.5 Is the fact that K is not recursive a *direct* consequence of Rice's Theorem? Justify your answer.

3.2.6 Let ψ_0, ψ_1, ψ_2, . . . be *any* list of partial functions from N to N. Suppose that the function $\psi(0) = 0$ and $\psi(x)$ undefined if $x \neq 0$ occurs in the list. Suppose also that for all i and j there is some e such that $\psi_i \circ \psi_j = \psi_e$. Prove that the characteristic function of the set $K = \{x : \psi_x(x)$ is defined$\}$ does not occur in the list ψ_0, ψ_1, ψ_2, ...

3.3 RECURSIVELY ENUMERABLE SETS

Even though the halting problem is algorithmically unsolvable, there is a sense in which it is "partially solvable." Specifically, there is a partial recursive function ψ which has value 0 on input x if x is in K and which is divergent if x is not in K. A program computing ψ on input x can be viewed as saying "yes" eventually if x is in fact in K, and never answering if x is not in K; such a program "partially solves" the halting problem, and is sometimes called a *partial decision procedure* for K. Intuitively, and computationally, this turns out to be equivalent to having a program which *lists* the elements of K : x is in K if and only if it eventually appears in the list generated by the program. Thus intuitively, a set is recursive if some program tests membership in the set, and we might call a set "recursively enumerable" if some program lists its members. Just as the notion of a set being recursive gives the most general possible condition for having an algorithm to decide what elements are in the set, it turns out that the notion of a set being recursively enumerable gives the most general possible condition for having an algorithm to list, or generate, the elements of the set. In this section we establish some fundamental facts about recursively enumerable sets, including an Extended Rice's Theorem which characterizes those properties of the input-output behavior of programs which are partially solvable.

Formally, we say that a set A is *recursively enumerable* (*r.e.*, for short) if A is the empty set or if A is the range of a total recursive function f, called an *enumerating function* for A; that is, if $A = \{f(0), f(1), f(2), . . .\}$. Notice that we have not required enumerating functions to be one-to-one; thus f may make repetitions while it is listing A. Our first proposition establishes a basic intuitive relationship between recursive and recursively enumerable sets.

3.3.1 PROPOSITION *A set is recursive if and only if both it and its complement are recursively enumerable.*

Proof Since the recursive sets are closed under complementation, establishing the "only if" part of the proposition amounts to showing that every recursive set is recursively enumerable. The empty set is r.e. by definition; suppose that A is a nonempty recursive set with $y \in A$, then

$$f(x) = \begin{cases} x & \text{if } c_A(x) = 1 \\ y & \text{if } c_A(x) = 0 \end{cases}$$

yields an enumerating function for A.

For the "if" part of the proposition, let A be an r.e. set whose complement is also r.e. If either A or its complement is empty, then certainly A is recursive. Otherwise, let f be an enumerating function for A and let g be an enumerating function for the complement of A. Then for all x,

$$c_A(x) = \begin{cases} 1 & \text{if } f(\min y[f(y) = x \text{ or } g(y) = x]) = x \\ 0 & \text{otherwise.} \end{cases}$$

In other words, to test whether or not x is in A, we list both A and its complement, simultaneously, until x appears on one of the lists and then we see which of the two lists x appeared on.   □

The next proposition gives us some more "liberal" conditions for defining and generating recursively enumerable sets.

3.3.2 PROPOSITION   *A set is recursively enumerable if and only if it is the range of a partial recursive function, and also if and only if it is the domain of a partial recursive function.*

Proof Every nonempty r.e. set is the range of a partial (in fact a total) recursive function by definition, and the empty set is the range of the totally undefined function. The other two implications which complete the proof of the proposition follow immediately from the next two lemmas.

3.3.3 LEMMA   *There is a total recursive function g such that for all x, the range of ϕ_x equals the domain of $\phi_{g(x)}$.*

Proof Let *step* be the step counting function given by Exercise 3.1.6, and define the partial recursive function θ by

$$\theta(x, y) = \min z[step(x, \Pi_1(z), \Pi_2(z)) = y + 1].$$

For any x and y, θ computes ϕ_x on all possible arguments $\Pi_1(z)$ for all possible numbers of steps $\Pi_2(z)$ to see whether y is in the range of ϕ_x. This type of computation is often referred to as a "dovetailing" computation in honor of Wood U. Dovetail, who first performed such a computation. Note that $\theta(x, y)$ is convergent if and only if y is in the range of ϕ_x. Now let i be such that $\theta = \phi_i$, and define $g(x) = s(i, 1, x)$ with s an s-m-n function. Then

$$\phi_{g(x)}(y) = \theta(x, y) = \begin{cases} \text{convergent} & \text{if } y \in \text{range of } \phi_x \\ \text{divergent} & \text{if } y \notin \text{range of } \phi_x \end{cases}$$

and hence the range of ϕ_x is the domain of $\phi_{g(x)}$. $\quad\square$

3.3.4 LEMMA *There is a total recursive function h such that if ϕ_x has a nonempty domain then $\phi_{h(x)}$ is an enumerating function for the domain of ϕ_x.*

Proof The proof is similar to that for the previous lemma. Define $\theta(x, 0) = \Pi_1(min\ z[step(x, \Pi_1(z), \Pi_2(z)) \neq 0])$ and

$$\theta(x, y + 1) = \begin{cases} \theta(x, y) & \text{if } step(x, \Pi_1(y + 1), \Pi_2(y + 1)) = 0 \\ \Pi_1(y + 1) & \text{if } step(x, \Pi_1(y + 1), \Pi_2(y + 1)) \neq 0. \end{cases}$$

If $\theta = \phi_i$ then $h(x) = s(i, 1, x)$ works.

3.3.5 EXERCISE Fill in the missing details for the proof of the previous lemma. $\quad\square$

To complete the proof of Proposition 3.3.2 we now observe that by Lemma 3.3.3 the range of every partial recursive function is the domain of some partial recursive function, and by Lemma 3.3.4 the domain of every partial recursive function is either empty or has an enumerating function, and hence is an r.e. set. $\quad\square$

Notice that Proposition 3.3.2 establishes the equivalence of our notions of recursively enumerable sets and of partially solvable problems; the class of r.e. sets is indeed the same as the class of domains of partial recursive functions.

Note also that the set K is the domain of the partial recursive function ψ such that $\psi(x) = \phi_x(x) = \phi_{univ}(x, x)$. Thus, the previous proposition provides a very simple proof that K is recursively enumerable. The next proposition is a formalized version of the argument at the beginning of Section 1.5 which shows that no effective list of (programs for) total recursive functions can include all of the total recursive functions.

3.3.6 PROPOSITION *Suppose that A is any recursively enumerable set such that ϕ_x is total for every x in A. There is a total recursive function g such that if $g = \phi_y$ then $y \notin A$.*

Proof If A is the empty set, then any total recursive g will do. If A is nonempty, let f be an enumerating function for A. Define

$$g(x) = \phi_{univ}(f(x), x) + 1 = \phi_{f(x)}(x) + 1.$$

Suppose, for the sake of a contradiction, that $g = \phi_y$ and $y \in A$. Then $y = f(z)$ for some z, and

$$\phi_y(z) = g(z) = \phi_{f(z)}(z) + 1 = \phi_y(z) + 1,$$

which is impossible. $\square$

We have introduced recursively enumerable sets partly as a way to deal with the notion of partially solvable problems, and the characterization in Proposition 3.3.2 makes it clear that r.e. sets do capture the intuitive notion of partially solvable problems. In the next theorem we characterize the kinds of input-output behavior of programs which are partially solvable in this sense. It is obvious that there are some nontrivial properties which are partially solvable; for example, the nonrecursive sets $B(y) = \{x : y$ is in the range of $\phi_x\}$ and $C(y,z) = \{x : \phi_x(y) = z\}$ of Proposition 3.2.2 are recursively enumerable.

3.3.7 EXERCISE   Show that the sets $B(y)$ and $C(y,z)$ of Proposition 3.2.2 are r.e.

Intuitively, these sets are r.e. because it is possible to run any program in an ongoing computation in such a way that if the program belongs in the set, this fact becomes obvious from the input-output behavior of the program after a finite amount of time. We shall see presently that the types of input-output behavior of programs which are partially solvable are just those types of behavior which can be observed after running programs in an appropriate manner for a finite amount of time.

In order to state the next theorem, we need a notation which enables us to talk about finite input-output behaviors, that is a notation for functions from finite subsets of N into N. Recall that in Exercise 2.1.12 we defined the primitive recursive function F by $F(x, y) = \Pi(y+1, \Pi_1(x)+1, \Pi_2(x))$ and we defined the sequence of partial functions $\pi_0, \pi_1, \ldots$ by

$$\pi_x(y) = \begin{cases} F(x, y) - 1 & \text{if } 0 < F(x, y) \text{ and } y < \Pi_1(x) + 1 \\ undefined & \text{otherwise.} \end{cases}$$

Then you showed that the sequence $\pi_0, \pi_1, \ldots$ contained all of the functions from finite subsets of N into N; we shall use this as our indexing of the finite functions.

3.3.8 EXTENDED RICE'S THEOREM   *Let $\mathscr{C}$ be any class of partial recursive functions. $P_\mathscr{C} = \{x : \phi_x \in \mathscr{C}\}$ is recursively enumerable if and only if there is a recursively enumerable set A such that*

$$\phi_x \in \mathscr{C} \qquad iff \qquad \pi_y \subseteq \phi_x \text{ for some } y \in A.$$

That is, $P_\mathscr{C}$ is r.e. if and only if some program can list a set of finite input-output behaviors such that the partial recursive functions in $\mathscr{C}$ are precisely those functions which "exhibit" at least one of the finite input-output behaviors on the list (expressed formally in terms of being extensions of one of the finite functions on the list).

Proof Proving the "if" part of the theorem is conceptually quite easy, but the notation gets a bit messy. Suppose that A is an r.e. set that meets the conditions of the theorem; this can be restated as the assertion that

$$P_\mathscr{C} = \{x : \pi_y \subseteq \phi_x \text{ for some } y \in A\}.$$

To see that then $P_\mathscr{C}$ is r.e. it is then sufficient to perform a dovetailing computation which generates all y's in A, and for each such y figures out what π_y is and then sees whether ϕ_x agrees with π_y on its (finite) domain; if and when such agreement is found, x is placed in $P_\mathscr{C}$. Formally, if A is the empty set, then so is $P_\mathscr{C}$ and we are done; otherwise let f be an enumerating function for A. If we define the partial function ψ such that

$$\psi(z) = \begin{cases} \Pi_1(z) & \text{if } \pi_{f(\Pi_2(z))} \subseteq \phi_{\Pi_1(z)} \\ divergent & \text{otherwise} \end{cases}$$

then $P_\mathscr{C}$ is certainly the range of ψ, and hence r.e. If you are a bit skeptical as to whether the definition of ψ really shows that it is partial recursive, we could expand the definition above to

$$\psi(z) = \begin{cases} \Pi_1(z) & \text{if } \forall w < \Pi_1(f(\Pi_2(z)))[F(f(\Pi_2(z)), w) > 0 \\ & \quad implies \ \phi_{\Pi_1(z)}(w) = F(f(\Pi_2(z)), w) - 1] \\ divergent & \text{otherwise,} \end{cases}$$

which should make everything perfectly clear!

To establish the "only if" part of the theorem, we use the following two lemmas:

3.3.9 LEMMA *If $P_\mathscr{C}$ is r.e. and $\phi \in \mathscr{C}$ then there is a $\pi_y \subseteq \phi$ such that $\pi_y \in \mathscr{C}$; that is, if $P_\mathscr{C}$ is r.e. then every partial function in $\mathscr{C}$ must have some finite subfunction also in $\mathscr{C}$.*

Proof Let $P_\mathscr{C} = $ domain of ϕ_i, let $\phi \in \mathscr{C}$, and let *step* be a step counting function. By our standard *s-m-n* construction, let g be a total recursive function such that

$$\phi_{g(x)}(y) = \begin{cases} \phi(y) & \text{if } \forall z \le y[step(x, x, z) = 0] \\ divergent & \text{if } \exists z \le y[step(x, x, z) \ne 0]. \end{cases}$$

3.3.10 EXERCISE Give an *s-m-n* construction which produces g.

Now notice that if $x \in K$ then $\phi_{g(x)}$ is a finite subfunction of ϕ, and if $x \notin$

K then $\phi_{g(x)} = \phi$. Since ϕ does belong to $\mathscr{C}$, if no finite subfunction of ϕ is in $\mathscr{C}$, then the complement of K is the domain of $\phi_i \circ g$. This would make the complement of K r.e. Along with the fact that K is r.e. and Proposition 3.3.1, this would make K recursive. Since K is not recursive, the lemma is proved.   □

3.3.11 LEMMA *If $P_{\mathscr{C}}$ is r.e., $\phi \in \mathscr{C}$, and ψ is a partial recursive function such that $\phi \subseteq \psi$ then $\psi \in \mathscr{C}$; that is, if $P_{\mathscr{C}}$ is r.e. then every partial recursive function which extends a function in $\mathscr{C}$ must also be in $\mathscr{C}$.*

Proof Let $P_{\mathscr{C}} = $ domain of ϕ_i and let ϕ and ψ be as stated. Using our standard *s-m-n* construction and a step counting function construction, let h be a total recursive function such that

$$\phi_{h(x)} = \begin{cases} \psi & \text{if } x \in K \\ \phi & \text{if } x \notin K. \end{cases}$$

Then if $\psi \notin \mathscr{C}$, the complement of K is the domain of $\phi_i \circ h$ and K is recursive as above.

3.3.12 EXERCISE Give a construction to produce the function h above; you may find this a bit more difficult than the previous exercise. (For a hint, see the restatement at the end of the section.)   □

We now return to the proof of the "only if" part of the theorem. Let $P_{\mathscr{C}} = $ domain of ϕ_i, and by our standard *s-m-n* construction let k be a total recursive function such that $\phi_{k(y)} = \pi_y$ for all y. Now define the r.e. set A to be the domain of $\phi_i \circ k$; then for all y, $y \in A$ if and only if $\pi_y \in \mathscr{C}$. Finally, using Lemmas 3.3.9 and 3.3.11 it is easy to see that

$$\phi_x \in \mathscr{C} \quad \text{iff} \quad \pi_y \subseteq \phi_x \text{ for some } y \in A.$$

which completes the proof of the Extended Rice's Theorem.   □

While the Extended Rice's Theorem does provide a complete characterization of the types of input-output behaviors of programs which are partially solvable, it is heavier artillery than is needed to settle most questions of this type which come up. In particular, when a set $P_{\mathscr{C}}$ is not r.e., this can often be shown by using the relatively simpler Lemma 3.3.9 or 3.3.11 rather than the full theorem.

3.3.13 EXERCISE

(a) Show that neither $\{x : \phi_x$ is the totally undefined function$\}$ nor $\{x :$ the domain of ϕ_x is finite$\}$ is r.e.

(b) Are either of the sets $\{x : \phi_x$ is *not* the totally undefined function$\}$ or $\{x :$ the domain of ϕ_x is infinite$\}$ r.e.? Prove your answer.

We conclude this section by observing an important consequence of the Extended Rice's Theorem. For any acceptable programming system $\phi_0, \phi_1, \ldots$ both the set $\{x : \phi_x$ is total$\}$ and its complement $\{x : \phi_x$ is not total$\}$ are not r.e. (The first violates Lemma 3.3.9, the second violates Lemma 3.3.11.) Therefore there is a sense in which the problem of deciding whether a program halts on all inputs is even "more unsolvable" than the halting problem. The halting problem is partially solvable, but the problem of whether programs halt on all inputs and its complement both fail to be even partially solvable. This is one reason we have deliberately not made the distinction between "procedures" and "algorithms" which is made by some authors. According to this distinction, a "procedure" is any program (or what *we* refer to as an algorithm) while an "algorithm" is a "procedure" which halts on all inputs. Because this distinction is so far from being algorithmically decidable, yielding a set of "algorithms" which is not even recursively enumerable, the distinction is inappropriate to a theory of algorithms. It is certainly part of good programming practice to try to make sure that the programs which we write always halt, and to write and document programs in such a way that others will be able to see this. Results such as the Extended Rice's Theorem make clear just how difficult a task this can be, and they underscore the need for computer science to develop systematic approaches to programming which facilitate accomplishing this and similar tasks.

Additional Exercises

*3.3.12 Construct h for the proof of Lemma 3.3.11. Intuitively, to compute $\phi_{h(x)}(y)$ one first begins computing $\phi(y)$ and simultaneously enumerating K. If $\phi(y)$ converges before x appears in K then $\phi_{h(x)}(y) = \phi(y)$. If x turns up in K before $\phi(y)$ converges, then one stops trying to compute $\phi(y)$ and begins instead computing $\psi(y)$. If and when $\psi(y)$ converges, one sets $\phi_{h(x)}(y) = \psi(y)$. How is the fact that $\phi \subseteq \psi$ used?

3.3.14 Show that a nonempty r.e. set is recursive if and only if it has a nondecreasing enumerating function.

3.3.15 Show that an infinite r.e. set is recursive if and only if it has a strictly increasing enumerating function.

3.3.16 Show that every infinite r.e. set has a one-to-one enumerating function.

3.3.17 Show that every infinite r.e. set has an r.e. subset which is not recursive.

3.3.18   Show that there are total recursive functions f and g such that

$$\text{domain } \phi_{f(x,y)} = \text{domain } \phi_x \cap \text{domain } \phi_y$$

and

$$\text{domain } \phi_{g(x,y)} = \text{domain } \phi_x \cup \text{domain } \phi_y.$$

(Thus the class of r.e. sets is closed under union and intersection. It is not closed under complementation. Why?)

3.3.19   For each of the two sets below, show whether or not it is recursive, whether or not it is r.e., and whether its complement is r.e.:

$A = \{x : \text{there are } y \text{ and } z \text{ such that } \phi_x(y) \text{ and } \phi_y(z) \text{ are convergent}\}$;
$B = \{x : \text{there is a } y \text{ such that } \phi_x(y) \text{ is convergent and } \phi_y \text{ is total}\}$.

3.3.20   Let $C = \{x : \phi_x(0) = 0\}$ and $D = \{x : \phi_x(0) = 1\}$. Show that there is no recursive set R such that $C \subseteq R$ and $D \cap R = \emptyset$; i.e., show that C and D are recursively inseparable. (See Exercises 2.4.13 and 3.1.9.)

3.4   THE RECURSION THEOREM AND ROGERS' ISOMORPHISM THEOREM

The Recursion Theorem is the "fixed point" theorem of the theory of the computable functions. It states that for any effective mapping of programs to programs, there is a "fixed point" program which is mapped to an equivalent program. The Recursion Theorem has many important applications, including the justification of general types of recursive definitions of functions, such as ALGOL-like recursive procedure definitions, in *any* acceptable programming system.

3.4.1 RECURSION THEOREM   *For every total recursive function f there is a natural number n (depending on f) such that*

$$\phi_n = \phi_{f(n)}.$$

Proof   There is a total recursive function g such that

$$\phi_{g(x)} = \begin{cases} \phi_{\phi_x(x)} & \text{if } \phi_x(x) \text{ is convergent} \\ \emptyset & \text{otherwise,} \end{cases}$$

where $\emptyset$ is, of course, the totally undefined function. Intuitively, the program $g(x)$ on input y first computes $\phi_x(x)$ and if and when this computation halts proceeds to compute $\phi_{\phi_x(x)}(y)$. To get the function g, define

$$\theta(x, y) = \phi_{univ}(\phi_{univ}(x, x), y)$$

and continue with our standard s-m-n construction. Let m be a program such that $f \circ g = \phi_m$, and let $n = g(m)$. Then since ϕ_m is total, $\phi_m(m)$ is convergent and we have that

$$\phi_n = \phi_{g(m)} = \phi_{\phi_m(m)} = \phi_{f(g(m))} = \phi_{f(n)}$$

Therefore, n is our required fixed point program.   □

As a very simple application of the Recursion Theorem we can show that there is an n such that ϕ_n is the constant function with output n. To do so we use our standard s-m-n construction to produce a total recursive function f such that $\phi_{f(x)}$ is the constant function x, and then we take n to be a fixed point for f. Such a program n might be called a "self-reproducing" program. (You might try to write a FORTRAN program which prints *itself*, and nothing else.) As another application of the Recursion Theorem, we can give a simple proof of Rice's Theorem. Let $\mathscr{C}$ be such that $\emptyset \neq P_\mathscr{C} \neq N$, and let $j \in P_\mathscr{C}$ and $k \notin P_\mathscr{C}$. Define $f(x) = k$ if $x \in P_\mathscr{C}$ and $f(x) = j$ if $x \notin P_\mathscr{C}$. If $P_\mathscr{C}$ were recursive then f would be a total recursive function which would not have a fixed point (since $x \in P_\mathscr{C}$ iff $f(x) \notin P_\mathscr{C}$ for all x), contradicting the Recursion Theorem.

The proof of the Recursion Theorem is actually somewhat stronger than the statement of the theorem. Although the Recursion Theorem we have stated is sufficient for most applications, there are some occasions when we need the stronger version below. The Extended Recursion Theorem asserts that fixed point programs can be found effectively from a program for the function f.

3.4.2 EXTENDED RECURSION THEOREM *There is a total recursive function n such that for all x, if ϕ_x is total, then*

$$\phi_{\phi_x(n(x))} = \phi_{n(x)} .$$

Proof Let i be a program for the function g in the proof of the Recursion Theorem; that is, $\phi_i = g$. Let c be a total recursive function for composition; that is, $\phi_{c(x,y)} = \phi_x \circ \phi_y$. Then $n(x) = g(c(x, i))$ is the required fixed point function.   □

The next proposition is an example of the use of the Recursion Theorem to justify a recursive definition of a function and it is an important result in its own right.

3.4.3 PROPOSITION *Let h be any total recursive function, and let s be an s-m-n function. There is an i such that $\phi_{h(x)} = \phi_{s(i,1,x)}$ for all x and such that $s(i, 1, x)$ is one-to-one as a function of x.*

Proof First, we claim that there is a program i such that

$$\phi_i(j, y) = \begin{cases} 0 & \text{if } \exists k < j[s(i, 1, k) = s(i, 1, j)] \\[2mm] 1 & \text{if } \forall k < j[s(i, 1, k) \neq s(i, 1, j)] \text{ and} \\ & \quad \exists k \leq y[k > j \text{ and } s(i, 1, k) = s(i, 1, j)] \\[2mm] \phi_{h(j)}(y) & \text{otherwise} \end{cases}$$

Notice that the program i has been defined "recursively" since it refers to itself on the right-hand side above. To get the program i, we first use an *s-m-n* construction to get a total recursive function f such that

$$\phi_{f(z)}(j, y) = \begin{cases} 0 & \text{if } \exists k < j[s(z, 1, k) = s(z, 1, j)] \\[2mm] 1 & \text{if } \forall k < j[s(z, 1, k) \neq s(z, 1, j)] \text{ and} \\ & \quad \exists k \leq y[k > j \text{ and } s(z, 1, k) = s(z, 1, j)] \\[2mm] \phi_{h(j)}(y) & \text{otherwise} \end{cases}$$

Then we apply the Recursion Theorem to get the required program i as a fixed point for the function f.

Assume, for the sake of a contradiction, that $s(i, 1, x)$ is not one-to-one as a function of x, and let j and k be least such that $s(i, 1, k) = s(i, 1, j)$ with $k < j$. Then for all y we have

$$\phi_{s(i,1,j)}(y) = \phi_i(j, y) = 0$$

and for all $y \geq j$ we have

$$\phi_{s(i,1,k)}(y) = \phi_i(k, y) = 1,$$

contradicting $s(i, 1, k) = s(i, 1, j)$. Therefore, $s(i, 1, x)$ must be one-to-one as a function of x. It then follows from the definition of i that

$$\phi_{s(i,1,x)}(y) = \phi_i(x, y) = \phi_{h(x)}(y)$$

for all x and y, which completes the proof of the proposition.   □

We now proceed to use this proposition to prove some important properties of acceptable programming systems. The first one extends Theorem 3.1.5 about effective translations to show that effective translations between acceptable programming systems can always be taken to be one-to-one.

3.4.4 PROPOSITION *Let* $\phi_0, \phi_1, \ldots$ *and* $\psi_0, \psi_1, \ldots$ *be any two acceptable programming systems. There is a one-to-one total recursive (translation) function f such that* $\phi_x = \psi_{f(x)}$ *for all* x.

Proof   Let g be a total recursive (translation) function such that $\phi_x = \psi_{g(x)}$ as given by Theorem 3.1.5. Let i be given by the previous proposition such that $s(i, 1, x)$ is one-to-one as a function of x and $\psi_{g(x)} = \psi_{s(i,1,x)}$. Then defining $f(x) = s(i, 1, x)$ we have the required one-to-one translation function from $\phi_0, \phi_1, \ldots$ to $\psi_0, \psi_1, \ldots$.   ☐

Recall Exercise 2.4.9 in which you showed that for the RAM programming system there is a one-to-one primitive recursive "padding" function p such that $\phi_x = \phi_{p(x,y)}$ for all x and y. Our next proposition shows that all acceptable programming systems have effective padding functions. This certainly coincides with our intuition that in any "reasonable" programming system it is easy to "pad" any program with "harmless" instructions to get a new, but equivalent, program. Notice that we could not transfer the padding function for the RAM system to other systems using the translations provided by Theorem 3.1.5, since they were not known to be one-to-one and hence might spoil the one-to-oneness of the padding function. Thus Proposition 3.4.4 is just what we need.

3.4.5 PROPOSITION (Padding Lemma)   *For any acceptable programming system $\psi_0, \psi_1, \ldots$ there is a one-to-one total recursive function p such that for all x and y, $\psi_x = \psi_{p(x,y)}$.*

Proof   Let $\phi_0, \phi_1, \ldots$ be our RAM programming system, and let p' be a padding function for that system as provided by Exercise 2.4.9. Let f and g be one-to-one total recursive functions such that $\phi_x = \psi_{f(x)}$ and $\psi_x = \phi_{g(x)}$ for all x. Then by defining $p(x,y) = f(p'(g(x),y))$ we are done.

3.4.6 EXERCISE   Verify that the function p defined above does indeed satisfy the conditions of the Padding Lemma.   ☐

Note that a careful inspection of the proof of the previous two propositions and the proof of the *s-m-n* Theorem shows that if the programming system $\psi_0, \psi_1, \ldots$ has a primitive recursive function for composition, then p is primitive recursive. Thus if an acceptable programming system has a primitive recursive function for composition it will also have primitive recursive *s-m-n* and padding functions, and translations of other programming systems into it can always be made by primitive recursive functions.

We close this section with the fundamental Rogers' Isomorphism Theorem which shows that not only can any two acceptable programming systems be translated effectively into each other, but that between any two acceptable programming systems there is an effective, one-to-one, and onto translation (that is, an "isomorphism"). Thus in a very *strong sense*, any two acceptable programming systems are equivalent.

3.4.7 ROGERS' ISOMORPHISM THEOREM *Let ϕ_0, ϕ_1, . . . and ψ_0, ψ_1, . . . be any two acceptable programming systems. There is a one-to-one, onto, total recursive function f such that $\phi_x = \psi_{f(x)}$ for all x.*

Proof For the proof, we need translations between the systems which are a bit nicer than those provided by Theorem 3.1.5 and Proposition 3.4.4. They are supplied by the following lemma.

3.4.8 LEMMA *There is a total recursive (translation) function g such that $\phi_x = \psi_{g(x)}$ and $0 < g(x) < g(x+1)$ for all x.*

Proof Intuitively, we simply take any translation and pad the translated programs until they are long enough. Formally, let h be the translation function given by Theorem 3.1.5, and define.

$$g(0) = p(h(0), \min y[p(h(0), y) > 0])$$

and

$$g(x + 1) = p(h(x + 1), \min y[p(h(x + 1), y) > g(x)])$$

where p is a padding function.   □

We now continue with the proof of Rogers' Isomorphism Theorem. Let g and h be total recursive functions such that $\phi_x = \psi_{g(x)}$, $\psi_x = \phi_{h(x)}$, $0 < g(x) < g(x+1)$, and $0 < h(x) < h(x+1)$ for all x. Notice that $g(x) > x$ and $h(x) > x$ for all x; this "strictly increasing" property is the key property of the translations g and h. Suppose that we take a program x in one of the programming systems, say ϕ_0, ϕ_1, . . . , and we begin "backtracking" from x under the *inverse* functions h^{-1} and g^{-1} of h and g. If x is in the range of h then $h^{-1}(x)$ exists, otherwise not. If $h^{-1}(x)$ exists and is in the range of g, then $g^{-1}(h^{-1}(x))$ exists and we can ask whether it is in the range of h. Because each time we apply one of our inverse functions the number we have decreases, this process cannot go on indefinitely; certainly by the time we apply the xth inverse, and perhaps a lot sooner, we must hit a "dead end" and get a program which is not in the range of the other translation. For our purposes, there are two distinct possibilities we are interested in: we "dead end" in the system ϕ_0, ϕ_1, . . . , or we "dead end" in the system ψ_0, ψ_1, ... Figure 3.4.9 illustrates these two situations.

If k is a function, let us define the notation k^n to stand for the n-fold composition of k with itself; that is,

$$k^0(x) = x, \quad k^1(x) = k(x), \quad k^2(x) = k(k(x)), \quad \text{and so on.}$$

Then the distinction we have just made can be expressed more formally by saying that either $(g^{-1} \circ h^{-1})^i(x)$ exists and is not in the range of h for

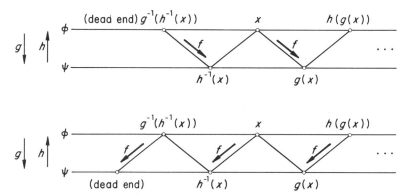

FIGURE 3.4.9   The two cases for the construction of the isomorphism f.

some i, or $h^{-1}((g^{-1}\circ h^{-1})^i(x))$ exists and is not in the range of g for some i. Now notice that in either case, *all* of the programs $(g^{-1}\circ h^{-1})^j(x)$ and $(h\circ g)^j(x)$ have the property of dead ending in the *same* system. Moreover, for any two programs x and y the two sets of programs

$$\{\ldots, (g^{-1}\circ h^{-1})^2(x), (g^{-1}\circ h^{-1})(x), x, (h\circ g)(x), (h\circ g)^2(x), \ldots\}$$

and

$$\{\ldots, (g^{-1}\circ h^{-1})^2(y), (g^{-1}\circ h^{-1})(y), y, (h\circ g)(y), (h\circ g)^2(y), \ldots\}$$

either are *the same* or are *disjoint*, since g and h are strictly increasing. Then if we define

$$f(x) = \begin{cases} g(x) & \text{if } x \text{ dead ends in } \phi_0, \phi_1, \ldots \\ h^{-1}(x) & \text{if } x \text{ dead ends in } \psi_0, \psi_1, \ldots, \end{cases}$$

f is a one-to-one and onto translation.

3.4.10 EXERCISE

(a) Verify that f is indeed a one-to-one and onto translation.
(b) Show that f is a total recursive function; that is, show how to compute f given algorithms for computing g and h.

This completes the proof of the Rogers' Isomorphism Theorem. The reader who is familiar with the Cantor-Bernstein Theorem of set theory will probably recognize that our construction of f uses two thirds of the standard proof of that theorem.   □

Additional Exercises

3.4.11 Show whether or not there are natural numbers m with each of the following properties:

(a) domain $\phi_m = \{m^2\}$

(b) domain $\phi_m = N - \{m\}$

(c) domain $\phi_m = \{x : \phi_m(x) \text{ is divergent}\}$

3.4.12 Show that there is a total recursive function f such that the set $\{i : \phi_i = \phi_{f(i)}\}$ is not recursive.

3.4.13

(a) Is there a natural number m such that domain $\phi_m = K$ and $m \in$ domain ϕ_m? Prove your answer.

(b) Is there a natural number n such that domain $\phi_n = K$ and $n \notin$ domain ϕ_n? Prove your answer.

*3.4.14 Show that if f is a one-to-one and onto total recursive function then so is f^{-1}. Conclude that if f is an isomorphism between two acceptable programming systems then f^{-1} is also an isomorphism between the systems, in the other order.

3.4.15 Consider a "natural" programming system such as our RAM system or Turing machines. In the proof of the Recursion Theorem, how will the "running time" of the fixed point program n be related to the running time of the program $f(n)$? Your first impulse might be to think that the program $f(n)$ calls the program n as a subroutine, and that its running time would be longer than that of n. But this is not the case; in fact the running time of $f(n)$ will often be less than that of n. Figure out a bound on just how much longer the running time of the program n can be than the running time of $f(n)$.

3.4.16 Use the Recursion Theorem to give a simple proof of Exercise 3.3.20.

3.4.17 Let $\mathscr{C}$ and $\mathscr{D}$ be any two disjoint properties of the input-output behavior of programs; that is, $\mathscr{C}$ and $\mathscr{D}$ are disjoint sets of partial recursive functions. Show that $P_{\mathscr{C}}$ and $P_{\mathscr{D}}$ are recursively inseparable; that is, show that there is no recursive set R such that $P_{\mathscr{C}} \subseteq R$ and $P_{\mathscr{D}} \cap R = \emptyset$. (Note that this is another generalization of Rice's Theorem.) (Clearly, $\mathscr{C}$ and $\mathscr{D}$ must be assumed nonempty!)

3.4.18 Let $\mathscr{C}$ and $\mathscr{D}$ be any two disjoint sets of partial recursive functions.

(a) Suppose that there are distinct partial functions ϕ and ψ such that $\phi \subseteq \psi$, $\phi \in \mathscr{C}$, and $\psi \in \mathscr{D}$. Show that there is no r.e. set S such that $P_{\mathscr{C}} \subseteq S$ and $P_{\mathscr{D}} \cap S = \emptyset$; in this case $P_{\mathscr{C}}$ is said to be *r.e. inseparable* from $P_{\mathscr{D}}$.

(b) Give examples of disjoint sets $\mathscr{C}$ and $\mathscr{D}$ such that $P_{\mathscr{C}}$ is r.e. inseparable from $P_{\mathscr{D}}$, but $P_{\mathscr{D}}$ is r.e. *separable* from $P_{\mathscr{C}}$.

*3.4.19 Extend Proposition 3.4.3 by proving that if h is any total recursive function then there exists a total recursive h' such that $\phi_{h(i)} = \phi_{h'(i)}$ for all i and $h'(x + 1) > h'(x)$ for all x. (Thus the range of h' is recursive.)

*3.4.20 Show that the Recursion Theorem (3.4.1) is equivalent to the following alternate version of the Recursion Theorem: for every partial recursive function ϕ_e there is a program i such that $\phi_i(x) = \phi_e(i, x)$ for all x.

Chapter 4

Applications to Mathematical Logic

The main goal of this (optional) chapter is to prove the fundamental Gödel Undecidability and Incompleteness Theorems. In 1931, Kurt Gödel published his celebrated paper on the incompleteness of formal arithmetic. This work had a profound effect on developments both in the foundations of mathematics and in the theory of computability. It has served as one of the principal contributions to our current understanding of the nature of mathematical truth, and it has provided much of the intuitive and technical basis for the theory of algorithms. There are four basic results which followed from this work of Gödel's.

1. Undecidability for Theorems: for any consistent theory at least as strong as arithmetic there is no effective procedure for determining whether or not an arbitrary sentence is a theorem. (A consistent theory is one which does not prove a contradiction.)
2. Incompleteness: for any consistent theory at least as strong as arithmetic there are sentences which can neither be proved nor disproved in the theory.[1]
3. Undecidability for Truth: there is no effective procedure for determining whether or not arbitrary sentences of arithmetic are true.
4. Unprovability of Consistency: for any consistent theory at least as strong as arithmetic, its consistency cannot be proved within the theory.[1]

The proof of the last of these requires the arithmetization (i.e., Gödel numbering) of a formal "proof theory" in order to formulate an arithmetical sentence which asserts "If I am provable then there is a shorter proof of my negation." Although the conceptual and general technical structure of this process is essentially the same as that involved in our indexing of RAM programs in Section 2.4, it nevertheless requires an excursion into mathematical logic which is beyond the scope of this book. However, using material on the partial recursive functions which we have already developed we can fairly easily give proofs of the first three of these results, and we can do this without

---

[1] For these results, we are of course considering only computationally reasonable theories—those for which the set of theorems can be enumerated effectively.

118

bothering to arithmetize a formal proof theory, or even bothering to *define* a formal proof theory. In addition to proving the Gödel Undecidability and Incompleteness Theorems, we prove another fundamental result of mathematical logic known as Church's Theorem.

5. Church's Theorem: there is no effective procedure for deciding whether or not an arbitrary mathematical sentence is logically valid (that is, true in all possible interpretations).

In this chapter we use two results about the partial recursive functions from earlier chapters. The first is Exercise 2.4.13 (redone in a general context in Exercise 3.1.9), which shows that if $A = \{x : \phi_x(x) = 0\}$ and $B = \{x : \phi_x(x) = 1\}$, then there is no recursive set R such that $A \subseteq R$ and $B \cap R = \emptyset$. Pairs of sets A and B with this property are called *recursively inseparable* since there is no effective procedure to separate one set from the other, even if we do not care what the procedure says about elements which are in neither set. The other result we use is the characterization of the partial recursive functions given in Section 2.3. Recall that in that section we showed that the class of functions gotten from addition, multiplication, the projection functions, and the equality predicate by using the operations of substitution and minimization—the class we called the min-computable functions—is exactly the same as the class of partial recursive functions. This result was given in Theorem 2.3.5.

4.1   A LANGUAGE FOR ARITHMETIC AND SOME AXIOMS

In preparation for our proofs of Gödel's results we start by defining $\mathscr{L}$, a *language for arithmetic*. $\mathscr{L}$ uses the following symbols:

1. variables: $x, y, z, \ldots$
2. constants: **0, 1**
3. functions: $\oplus, \otimes$
4. equality: $=$
5. logical connectives: $\sim, \vee, \&, \Rightarrow, \Leftrightarrow$
6. quantifiers: $\exists, \forall$
7. punctuation: (,)

We wish to emphasize that **0, 1**, $\oplus$, and $\otimes$ are *formal symbols* in the language $\mathscr{L}$, and they are *not necessarily* to be interpreted as standing for zero, one, addition, and multiplication, respectively. The *terms* of $\mathscr{L}$

are defined inductively as follows:

1. Variables and constants are terms.
2. If s and t are terms, so are $(s \oplus t)$ and $(s \otimes t)$.
3. Something is a term only by virtue of (1) and (2).

The *formulas* of $\mathscr{L}$ are defined inductively as follows:

1. If s and t are terms, $(s = t)$ is a formula (usually called an *atomic formula*).
2. If F and G are formulas then so are $\sim F$, $(F \vee G)$, $(F \mathbin{\&} G)$, $(F \Rightarrow G)$, $(F \Leftrightarrow G)$, $\exists xF$, and $\forall xF$, for all variables x.
3. Something is a formula only by virtue of 1 and 2.

In the formulas $\exists xF$ and $\forall xF$, F is called the *scope* of the x-quantifier. Intuitively, F is asserting something about x and $\forall xF$ asserts that it holds for all values of x. An occurrence of a variable x in a formula is said to be *bound* if it is in the scope of an x-quantifier; otherwise the occurrence is said to be *free*. A formula with no free (occurrences of) variables is said to be a *sentence*; sentences make statements which under any *given interpretation* are either true or false.

We now consider some examples to illustrate the definitions we have just given. $(y \oplus 1)$ is a term, and $(x = (y \oplus 1))$ and $\exists y(x = (y \oplus 1))$ are formulas; call them F and G, respectively. In G, F is the scope of the y-quantifier, and so in G the occurrence of y is bound and the occurrence of x is free; thus G is not a sentence. In the formula $(\forall xG \vee (x = 1))$, G is the scope of the x-quantifier and so the occurrence of x within G is bound, but the rightmost occurrence of x is free; we still do not have a sentence. It is perfectly permissible under our definition of $\mathscr{L}$ to have formulas such as this which have both bound and free occurrences of the same variable, but doing so is generally considered poor mathematical practice; we shall try to avoid it in the formulas we use. Finally, in this example, the formula $\forall xG$ is a sentence.

The sentence $\forall xG$—that is, the sentence $\forall x \exists y(x = (y \oplus 1))$—is true if we interpret the variables x and y as ranging over the set Z of all integers, $Z = \{\ldots -2, -1, 0, 1, 2, \ldots\}$, interpret $\oplus$ as addition on Z, and interpret 1 as the integer one. However, the sentence is false if we change the interpretation to have the variables range over $N = \{0, 1, 2, \ldots\}$, interpret $\oplus$ as addition on N, and interpret 1 as the integer one.

4.1.1 EXERCISE

(a) If we interpret the variables as ranging over N, $\oplus$ as multiplication on N, and 1 as the integer zero, is the sentence $\forall x \exists y(x = (y \oplus 1))$ true or false?

(b) If we interpret the variables as ranging over N, $\oplus$ as multiplication on N, and 1 as the integer one, is the sentence $\forall x \exists y(x = (y \oplus 1))$ true or false?

Unlike the sentence $\forall x \exists y(x=(y \oplus 1))$, the sentence $\forall x \exists y(y=(x \oplus 1))$ is true in *every* possible interpretation; no matter what (nonempty) set D the variables range over, no matter what function on D $\oplus$ is interpreted to be, and no matter what element of D 1 is interpreted to be, the sentence is true. Sentences like this which are true in every possible interpretation are called *logically valid*. On the other hand, the sentence $\exists x \sim (x = x)$ is false in every possible interpretation, and such sentences are called *logically false*.

4.1.2 EXERCISE Explain why a sentence S is logically false if and only if the sentence $\sim S$ is logically valid.

Since we are interested in developing a theory of arithmetic, the *usual* (or *standard*) *interpretation* of sentences of $\mathcal{L}$ will be to interpret the variables as ranging over the natural numbers, N, to interpret 0 as zero and 1 as one, and to interpret $\oplus$ as addition on N and $\otimes$ as multiplication on N. On the other hand, we are going to list some axioms for a theory of arithmetic, and the reader must be very careful *not to assume* that just because a sentence is true in the usual interpretation that it can be *proved* from the axioms. Indeed, one of the major results of this chapter is that no reasonable theory of arithmetic can prove all of the true sentences of arithmetic (that is, all of the sentences true in the usual interpretation).

We have been talking a great deal about sentences being true or false in given interpretations, and you may be wondering why we have not given a formal definition of what it means for a sentence to be true in a given interpretation. We have not given one because none is *necessary*; we all know quite well what it means when we say that some simple sentence is true in some given interpretation. Actually, it is quite easy to give a precise mathematical definition of what it means for an arbitrary sentence to be true in an interpretation; such a definition simply follows the inductive definition of $\mathcal{L}$ in the obvious ways. The interested reader can find such a definition in almost any standard textbook on mathematical logic, but we repeat that for our purposes we have no need for such a formal definition. We caution the reader not to make the mistake of thinking that knowing what it *means* for a sentence to be true in a given interpretation implies that there is an effective way of *determining* which sentences are true in that interpretation. We shall see later that there is no algorithm for determining which sentences of $\mathcal{L}$ are true in the usual interpretation, and this is one of the major points of this chapter: knowing what it *means* for something to be true is easy, but knowing what the truth *is* is very difficult. Of course, in spite of the fact that there is no general algorithm for determining truth, in many simple cases, such as the specific examples we have given so far in this section, it is quite easy to determine whether such simple sentences are true in a given interpretation. This is all we ask of the reader.

Similarly, when we *prove* sentences of $\mathscr{L}$ from the axioms we are about to give we shall proceed, as is almost always the case in mathematics, to give an argument which shows that the sentence being "proved" must be true in any interpretation in which all of the given axioms are true. This simply is standard mathematical practice. For example, when one proves a theorem of group theory, one gives an argument that for any interpretation in which the axioms are true (that is, in any *group*), the theorem must also be true. Also for example, all of the logically valid sentences one can write in the language of a theory must be theorems of the theory since, *whatever* the axioms of the theory are, whenever the axioms are true in an interpretation, the logically valid sentences will (trivially) also be true in the interpretation.

In order to simplify notation, for the rest of this chapter we shall not play strictly by the rules when writing sentences of $\mathscr{L}$, our language for arithmetic. For example, we follow the standard practice of omitting parentheses when there is no danger of ambiguity, and we use square brackets instead of parentheses when this helps with readability. We write the sentence $\forall x_1 \ldots \forall x_n F$ simply as F, with the understanding that all the free variables of F are universally quantified. We shall also use some abbreviations: $x \leq y$ stands for $\exists z(x \oplus z = y)$, $x < y$ stands for $x \leq y$ & $\sim(x = y)$, and $x \neq y$ stands for $\sim(x = y)$. Also, for each natural number n, if $m = n + 1$ then **m** stands for the term $(\mathbf{n} \oplus \mathbf{1})$. Note that this is an inductive definition of m. For example, **3** stands for the term $(((0 \oplus 1) \oplus 1) \oplus 1)$. Notice that we have committed the blunder of having **1** stand for both the *constant symbol* **1** and the *term* $(0 \oplus 1)$. Fortunately, axiom S1 below will rescue us from this predicament by declaring these two to be identical.

4.1.3 EXERCISE   Verify that on the basis of *logic* alone, that is on the basis of the properties of the logical connectives $\sim$, $\lor$, etc., that the sentence $x \leq y \Rightarrow (x < y \lor x = y)$ is "true"—that is, it is logically valid. Verify that the converse implication, $(x < y \lor x = y) \Rightarrow x \leq y$ is *not* logically valid.

The next definition lists some axioms for a very minimal theory of arithmetic.

4.1.4 DEFINITION   The following are our *axioms* for arithmetic:

S1. $0 \oplus 1 = 1$	A1. $x \oplus 0 = x$
S2. $x \oplus 1 \neq 0$	A2. $x \oplus (y \oplus 1) = (x \oplus y) \oplus 1$
S3. $x \neq 0 \Rightarrow \exists z(z \oplus 1 = x)$	M1. $x \otimes 0 = 0$
S4. $x \oplus 1 = y \oplus 1 \Rightarrow x = y$	M2. $x \otimes (y \oplus 1) = (x \otimes y) \oplus x$

$$\text{T1.} \quad x < y \;\; \vee \;\; x = y \;\; \vee \;\; y < x$$

Notice that the S-axioms give properties of "$\oplus 1$" as a successor function: S4 says it is one-to-one; S2 and S3 say it is "almost" onto; and S1 gives a "boundary condition." The A-axioms give the relation between the successor function and addition; that is, they give the usual recursive definition of addition. The M-axioms give the relation between addition and multiplication; that is, they give the usual recursive definition of multiplication. The T-axiom gives trichotomy.

4.1.5 EXERCISE   Give an interpretation in which the variables range over N and 0 and 1 are interpreted as 0 and 1, respectively, such that axioms S1 through S4 are true but "$\oplus 1$" is *not* the successor function; that is, give an "unusual" definition of addition as an interpretation for $\oplus$ which still satisfies the S-axioms.

We now wish to begin showing that certain sentences of $\mathscr{L}$ are provable from our axioms. Although it is possible, and not very difficult, to give a formal set of "rules of inference" which make it possible to decide what is and what is not a (valid) proof from our axioms, and also make it possible to generate all of the (valid) proofs, we reiterate that this formality is not necessary. Just as programmers can recognize many algorithms to be algorithms without a formal definition of what is and what is not an algorithm, we can recognize simple instances of mathematically correct proofs without a precise and rigorous definition of what a proof is. Actually, mathematicians and students of mathematics seldom, if ever, rely on *formal* definitions of "proof" in order to recognize valid proofs, and we are simply going to follow this standard practice. (In fact, the proofs we give actually use only extremely simple and obvious rules of inference; for example: from S and $S \Rightarrow T$ conclude T; from S and T conclude $S \,\&\, T$; from $S(\mathbf{n})$ conclude $\exists x S(x)$; and from $\forall x S(x)$ conclude $S(\mathbf{n})$ for the constant term $\mathbf{n}$.) We also use some obvious properties of functions and equality; that is, we use the fact that in any interpretation the function symbols $\oplus$ and $\otimes$ are always interpreted as standing for (total) functions (of two arguments) and the symbol "$=$" is always interpreted as standing for equality. For example, for any x and y there is always exactly one element which is the interpretation of $x \oplus y$, we may substitute equals for equals, and so on.

In much of what follows, we shall be proving that some infinite collection of sentences are all provable from our axioms. For example, in Proposition 4.1.6, we show that $x < \mathbf{n} \oplus 1 \Rightarrow x \leq \mathbf{n}$ is provable from the axioms for each n in N. Frequently, as in Proposition 4.1.7, it will be necessary to use mathematical induction to show that some infinite collections of sentences are all provable from our axioms. Note that such propositions are often called "metatheorems"—that is theorems *about* what is provable from our axioms. The use of mathematical induction is a standard mathematical practice, and we often use it to establish such "metatheorems" about what is provable from our axioms. We caution the reader that we are not permitted to use induction *within* our theory of arithmetic; that is, we may not use mathematical induction to prove a particular single sentence from our axioms, since mathematical induction has (deliberately) not been built into our axioms. Exercises 4.1.9 and 4.1.10 and the discussion preceding them should help to clarify this distinction.

We use the notation $\vdash S$ to denote that the sentence S is provable from our axioms. We caution the reader that many books on mathematical logic use this notation to indicate that there is a formal proof of S according to some given formal definition of proofs. Our meaning is simply that the reader, and anyone else with sufficient mathematical background, can recognize that S is provable from our axioms in the usual mathematical sense which we have just been discussing. In fact, using *any* reasonable formal definition of proofs, all of the proofs of sentences of $\mathcal{L}$ we give in this chapter are so simple that they could easily be translated into formal proofs satisfying such a formal definition.

We conclude this section by showing that some simple sentences are provable from our axioms.

4.1.6 PROPOSITION   *For all natural numbers n:* $\vdash x < \mathbf{n} \oplus 1 \Rightarrow x \leq \mathbf{n}.$

Proof $x < \mathbf{n} \oplus 1$ abbreviates $\exists z(x \oplus z = \mathbf{n} \oplus 1)$ & $x \neq \mathbf{n} \oplus 1$. If $z = 0$ then from A1 we get $x = \mathbf{n} \oplus 1$ and a contradiction; therefore $z \neq \mathbf{0}$. From S3 we then get $\exists w(x \oplus (w \oplus 1) = \mathbf{n} \oplus 1)$, and then A2 and S4 give us $\exists w(x \oplus w = \mathbf{n})$ which is abbreviated $x \leq \mathbf{n}$. Therefore $\vdash x < \mathbf{n} \oplus 1 \Rightarrow x \leq \mathbf{n}$.   □

4.1.7 PROPOSITION   *For all natural numbers m, n, and p:*

1. $\vdash x \leq \mathbf{n} \;\Rightarrow\; x = \mathbf{0} \bigvee x = \mathbf{1} \bigvee \ldots \bigvee x = \mathbf{n}\,;$
2. $m+n=p$ *implies* $\vdash \mathbf{m} \oplus \mathbf{n} = \mathbf{p}\,;$
3. $m{\cdot}n=p$ *implies* $\vdash \mathbf{m} \otimes \mathbf{n} = \mathbf{p}\,;$
4. $m \neq n$ *implies* $\vdash \mathbf{m} \neq \mathbf{n}.$

Proof

1. This part is proved by induction on n; we do it in complete detail. If $n = 0$ we want $\vdash x \leq 0 \Rightarrow x = 0$. Suppose $x \leq 0$; this abbreviates $\exists z(x \oplus z = 0)$. If $z \neq 0$ then by S3 $\exists w(x \oplus (w \oplus 1) = 0)$, and then by A2 $\exists w((x \oplus w) \oplus 1 = 0)$ which contradicts S2. Therefore $z = 0$, and we have $x \oplus 0 = 0$. Then by A1 we get $x = 0$. Thus $\vdash x \leq 0 \Rightarrow x = 0$.

   Now assume $\vdash x \leq \mathbf{n} \Rightarrow x = 0 \vee \ldots \vee x = \mathbf{n}$ as an induction hypothesis. Let $s = n+1$. Suppose $x \leq \mathbf{s}$; then since $\mathbf{s}$ is $\mathbf{n} \oplus 1$ we have $x < \mathbf{n} \oplus 1 \vee x = \mathbf{n} \oplus 1$ by Exercise 4.1.3. By Proposition 4.1.6 above we have $x \leq \mathbf{n} \vee x = \mathbf{n} \oplus 1$, and using our induction hypothesis we now get $x = 0 \vee \ldots \vee x = \mathbf{n} \vee x = \mathbf{n} \oplus 1$. Therefore,

   $$\vdash x \leq \mathbf{s} \quad \Rightarrow \quad x = 0 \vee \ldots \vee x = \mathbf{s},$$

   which completes the induction step and the proof of part 1.

2. We prove this part also by induction on n. If $n=0$ then $\vdash \mathbf{m} \oplus 0 = \mathbf{m}$ by A1. Let $m+n=p$, $s=n+1$, and $q=p+1$. Then $m+s=q$ and by induction hypothesis $\vdash \mathbf{m} \oplus \mathbf{n} = \mathbf{p}$. Thus $\vdash (\mathbf{m} \oplus \mathbf{n}) \oplus 1 = \mathbf{p} \oplus 1$ and so by A2 we have $\vdash \mathbf{m} \oplus (\mathbf{n} \oplus 1) = \mathbf{p} \oplus 1$, which by our abbreviations is just $\vdash \mathbf{m} \oplus \mathbf{s} = \mathbf{q}$. This completes the induction step and the proof of part 2.

3. Again, by induction on n. If $n=0$ then $\vdash \mathbf{m} \otimes 0 = 0$ by M1. Let $m \cdot n=p$, $s=n+1$, and $q=p+m$; then $m \cdot s=q$. By induction hypothesis $\vdash \mathbf{m} \otimes \mathbf{n} = \mathbf{p}$, and therefore $\vdash (\mathbf{m} \otimes \mathbf{n}) \oplus \mathbf{m} = \mathbf{p} \oplus \mathbf{m}$. Then by M2 we have $\vdash \mathbf{m} \otimes (\mathbf{n} \oplus 1) = \mathbf{p} \oplus \mathbf{m}$. Finally, by part 2 we have $\vdash \mathbf{p} \oplus \mathbf{m} = \mathbf{q}$ and since $\mathbf{s}$ is $\mathbf{n} \oplus 1$, $\vdash \mathbf{m} \otimes \mathbf{s} = \mathbf{q}$. This completes the proof of part 3. We leave part 4 as an exercise.

4.1.8 EXERCISE   Show that for all natural numbers m and n, if $m \neq n$ then $\vdash \mathbf{m} \neq \mathbf{n}$.   □

When we referred to our axioms as giving a very minimal theory of arithmetic, one of the things we meant was that they are obviously too weak in the sense that there are some very simple sentences of $\mathcal{L}$ which are true in the usual interpretation but which are not provable from our axioms. It was our desire to include in our axioms only what was necessary for establishing the results of this chapter, and nothing more. One of the main results of this chapter, the Gödel Incompleteness Theorem, shows that no reasonable theory of arithmetic, no matter how sophisticated, can prove all the sentences of $\mathcal{L}$ which are true in the usual interpretation. Indeed, developing this distinction between provability and truth is one of the principal goals of this chapter. This is illustrated by the next exercise. First notice that Proposition 4.1.7 part 4 shows, as a special case, that for each natural number n, $\vdash \mathbf{n} \neq \mathbf{n} \oplus \mathbf{1}$.

4.1.9 EXERCISE. *Try* to show $\vdash \forall x(x \neq x \oplus 1)$, and if you think you have succeeded go back and find your mistake.

The next exercise is probably fairly difficult, but we encourage the reader to attempt its solution in order to gain a fuller understanding of the preceding discussion.

4.1.10 EXERCISE   Give an interpretation in which all of the axioms in Definition 4.1.4 are true but in which the sentence $\forall x(x \neq x \oplus 1)$ is false.

4.2   REPRESENTABILITY OF PARTIAL RECURSIVE FUNCTIONS

Although our theory of arithmetic as given by our axioms in Definition 4.1.4 is very weak (as we discussed at the end of the last section), our axioms are nevertheless powerful enough to prove quite a wealth of arithmetical facts. Our goal in this section is to show that for any partial recursive function ϕ and for any pair of integers i and j, if $\phi(i) = j$ then this fact is provable from our axioms. To this end, we say that a partial function $\phi: N^k \to N$ is *representable* from our axioms if there is a formula $F_\phi(x_1, \ldots, x_k, z)$ of $\mathscr{L}$, where the free variables in F_ϕ are among $x_1, \ldots, x_k, z$, such that

$$\vdash F_\phi(x_1, \ldots, x_k, y) \,\&\, F_\phi(x_1, \ldots, x_k, z) \;\Rightarrow\; y = z;$$

and such that for all natural numbers $n_1, \ldots, n_k, m$, if

$$\phi(n_1, \ldots, n_k) = m$$

then

$$\vdash F_\phi(\mathbf{n}_1, \ldots, \mathbf{n}_k, \mathbf{m}).$$

Notice that the first condition says that whenever $F_\phi(x_1, \ldots, x_k, z)$ holds, then the value of z is *uniquely* determined; thus F_ϕ can be viewed as determining a partial function of the arguments $x_1, \ldots, x_k$. The second condition says that the partial function determined by F_ϕ agrees with ϕ on those arguments on which ϕ is defined.

In the next three propositions we show that all partial recursive functions are representable from our axioms; for this purpose we use the characterization of the partial recursive functions as the min-computable functions provided by Theorem 2.3.5. The proofs of the first two propositions are completely straightforward, and we leave parts of them as exercises. The proof of the third proposition is only slightly tricky, and we present all of it.

4.2.1 PROPOSITION *Addition, multiplication, the projection func-tions, and the characteristic function of the equality predicate are all representable from our axioms.*

Proof Addition is represented by the formula $x \oplus y = z$. To verify this we first observe that by the obvious properties of functions and equality we have

$$\vdash (x \oplus y = z) \,\&\, (x \oplus y = w) \;\Rightarrow\; z = w.$$

Notice that since this formula is logically valid we did not need any of the axioms to establish this! To finish, we also need to know that if $m + n = p$ then $\vdash \mathbf{m} \oplus \mathbf{n} = \mathbf{p}$. This was established in Proposition 4.1.7.

Multiplication is represented by the formula $x \otimes y = z$; the projection function P_i^k is represented by $x_i = z$; and the characteristic function of the equality predicate is represented by $(x = y \,\&\, z = \mathbf{1}) \vee (x \neq y \,\&\, z = \mathbf{0})$. Showing that these formulas work is very similar to showing that addition is representable, and is left as an exercise.

4.2.2 EXERCISE Complete the proof of Proposition 4.2.1 using Prop-osition 4.1.7. □

4.2.3 PROPOSITION *A partial function gotten by substitution from representable partial functions is itself representable. Specifically, let $\vec{x}$ stand for $x_1, \ldots, x_n$ and suppose that*

$$\phi(\vec{x}) = \psi(\theta_1(\vec{x}), \ldots, \theta_m(\vec{x}));$$

also suppose that $G_i(\vec{x}, y_i)$ represents θ_i for $1 \leq i \leq m$ and that $H(y_1, \ldots, y_m, z)$ represents ψ. Then

$$\exists y_1 \ldots \exists y_m [G_1(\vec{x}, y_1) \,\&\, \ldots \,\&\, G_m(\vec{x}, y_m) \,\&\, H(y_1, \ldots, y_m, z)]$$

represents ϕ.

Proof Let $k_1, \ldots, k_n$ be natural numbers and let $\vec{k}$ stand for $k_1, \ldots, k_n$. If $\phi(\vec{k}) = p$ then there are $q_1, \ldots, q_m$ such that $\theta_i(\vec{k}) = q_i$ for $1 \leq i \leq m$ and $\psi(q_1, \ldots, q_m) = p$. Then by our representability assumptions for $\theta_1, \ldots, \theta_m$ and ψ, $\vdash G_i(\mathbf{k}, \mathbf{q}_i)$ for $1 \leq i \leq m$ and $\vdash H(\mathbf{q}_1, \ldots, \mathbf{q}_m, \mathbf{p})$, and so

$$\vdash \exists y_1 \ldots \exists y_m [G_1(\vec{\mathbf{k}}, y_1) \,\&\, \ldots \,\&\, G_m(\vec{\mathbf{k}}, y_m) \,\&\, H(y_1, \ldots, y_m, \mathbf{p})].$$

The proof of the uniqueness statement is nearly as simple, and is left as an exercise.

4.2.4 EXERCISE Complete the proof of Proposition 4.2.3. □

4.2.5 PROPOSITION *The minimization of a representable function is* *representable. Specifically, let $\vec{x}$ stand for $x_1, \ldots, x_n$ and suppose that*

$$\phi(\vec{x}) = \min z[\psi(\vec{x}, z) = 0];$$

also suppose that $G(\vec{x}, z, y)$ represents ψ. Then ϕ is represented by the formula $F(\vec{x}, z)$ which is

$$\forall w(w \leq z \Rightarrow \exists y[G(\vec{x}, w, y) \,\&\, (y = 0 \Leftrightarrow w = z)]).$$

Proof   Note that $\vdash F(\vec{x}, z) \Rightarrow G(\vec{x}, z, 0)$, and from the definition of F

$$\vdash (w < z \,\&\, F(\vec{x}, z)) \Rightarrow \sim G(\vec{x}, w, 0),$$

and therefore

$$\vdash (w < z \,\&\, F(\vec{x}, z)) \Rightarrow \sim F(\vec{x}, w).$$

Then by axiom T1 we have

$$\vdash F(\vec{x}, z) \,\&\, F(\vec{x}, w) \Rightarrow z = w.$$

Let $k_1, \ldots, k_n$ be natural numbers and let $\vec{k}$ stand for $k_1, \ldots, k_n$. Suppose that $\phi(\vec{k}) = p$; then $\psi(\vec{k}, p) = 0$ and for all $q < p$, $\psi(\vec{k}, q) > 0$. Thus by our representability assumption for ψ, $\vdash G(\mathbf{k}, \mathbf{p}, 0)$ and for each $q < p$, $\vdash G(\mathbf{k}, \mathbf{q}, \mathbf{m})$ for some $m \neq 0$; $\vdash \mathbf{m} \neq 0$ by Proposition 4.1.7 part 4, and so from the uniqueness statement for G we get $\vdash \sim G(\mathbf{k}, \mathbf{q}, 0)$ for each $q < p$. If $p = 0$ then $w < \mathbf{p}$ contradicts the axioms (by Proposition 4.1.7 part 1 and the definition of $<$) and so certainly

$$\vdash \forall w(w < \mathbf{p} \Rightarrow \sim G(\mathbf{k}, w, 0)).$$

If $p = s + 1$ then by Propositions 4.1.6 and 4.1.7 part 1

$$\vdash w < \mathbf{p} \Rightarrow w = 0 \bigvee \ldots \bigvee w = \mathbf{s},$$

and so we also have

$$\vdash \forall w(w < \mathbf{p} \Rightarrow \sim G(\mathbf{k}, w, 0)).$$

Putting all of this together we have $\vdash F(\mathbf{k}, \mathbf{p})$, and the proof of the proposition is complete. $\square$

The previous three propositions, together with Theorem 2.3.5, prove the following theorem:

4.2.6 THEOREM   *Every partial recursive function is representable from the axioms given in Definition 4.1.4. Moreover there is an effective translation from the definitions (or programs) for the min-computable functions, ϕ, into the formulas, F_ϕ, of $\mathscr{L}$ which represent the functions.*

Additional Exercises

4.2.7 Prove that the translation F of Theorem 4.2.6 is even stronger than required in the sense that not only is it true that if $\phi(n_1, \ldots, n_k) = m$ then $\vdash F_\phi(\mathbf{n}_1, \ldots, \mathbf{n}_k, \mathbf{m})$ but in fact $\phi(n_1, \ldots, n_k) = m$ if and only if $\vdash F_\phi(\mathbf{n}_1, \ldots, \mathbf{n}_k, \mathbf{m})$.

4.2.8 Assuming that there is some algorithm for enumerating the theorems provable from the axioms of Definition 4.1.4, give an intuitive explanation of how to use any formula F for which

$$\vdash F(\vec{x}, y)\, \&\, F(\vec{x}, z) \;\Rightarrow\; y = z$$

as a "program" for computing a partial function represented by F. Thus, Propositions 4.2.1, 4.2.3, and 4.2.5 may be thought of as simply translating the MIN-programs of Section 2.3 into a new programming system.

4.3 UNDECIDABILITY AND INCOMPLETENESS THEOREMS

In this section we prove the Gödel Undecidability and Incompleteness Theorems as well as Church's Theorem. To prove these fundamental theorems, we use the results of the previous section on the representability of partial recursive functions (specifically, Theorem 4.2.6) together with the recursive inseparability of the two sets $S_0 = \{i : \phi_i(i) = 0\}$ and $S_1 = \{i : \phi_i(i) = 1\}$ (see Exercise 2.4.13 or 3.1.9). The ideas we use to prove these fundamental theorems are quite simple, and their proofs are very similar; in fact, we prove a single, general undecidability theorem and then deduce all of the fundamental theorems from it as corollaries.

First, we introduce some notation used throughout the rest of this section. Let ϕ_{univ} be a universal partial recursive function and let $F(x,y,z)$ be a formula of $\mathscr{L}$ which represents ϕ_{univ}. Of course, the existence of F is guaranteed by Theorem 4.2.6. Let $S(x)$ be the formula $F(x, x, 0)$; thus for each natural number i, $S(\mathbf{i})$ is a *sentence* which (in interpretations in which our axioms are true) "asserts" that $\phi_i(i) = 0$. Finally, let AX be the sentence S1 & S2 & . . . & M2 & T1, that is, the conjunction of all of our axioms from Definition 4.1.4. Thus, AX is simply a single sentence which is equivalent to our finite set of axioms.

Notice that since ϕ_{univ} is a universal function, $\phi_i(j) = \phi_{univ}(i, j)$. Since F represents ϕ_{univ}, if $\phi_i(i) = 0$ then $\vdash F(\mathbf{i}, \mathbf{i}, \mathbf{0})$; that is, if $\phi_i(i) = 0$ then $S(\mathbf{i})$ is provable from our axioms. Similarly, if $\phi_i(i) = 1$ then $\vdash F(\mathbf{i}, \mathbf{i}, \mathbf{1})$ and $\vdash (F(\mathbf{i}, \mathbf{i}, w) \Rightarrow w = \mathbf{1})$, and since $\vdash \mathbf{0} \neq \mathbf{1}$ we have that $\vdash \sim F(\mathbf{i}, \mathbf{i}, \mathbf{0})$; that is, if $\phi_i(i) = 1$ then $\sim S(\mathbf{i})$ is provable from AX. Thus if $\phi_i(i) = 1$ then the sentence $[AX\,\&\,S(\mathbf{i})]$ is logically false; that is, in any interpretation in which AX is true $S(i)$ is false, so $[AX \& S(i)]$ is false in *all* possible interpretations.

We need one more definition before we proceed with our general undecidability theorem. Let us call a mathematical theory *minimally adequate* if it contains (at least) all of the symbols of our language $\mathcal{L}$ for arithmetic, and if AX is a theorem of the theory; thus $S(i)$ is a theorem of the theory whenever $\phi_i(i) = 0$. Recall that a theory is consistent if it does not prove a contradiction. There are very many minimally adequate theories, including many *consistent* ones, which prove sentences which are *false* in the usual interpretation; see Exercise 4.3.8. Of course, we are most interested in those minimally adequate theories (such as those given by our axioms) for which every provable sentence of $\mathcal{L}$ happens to be true in the usual interpretation.

4.3.1 GENERAL UNDECIDABILITY THEOREM *For any minimally adequate theory, there is no algorithm for recursively separating the theorems of the theory from the logically false sentences.*

Proof If $\phi_i(i) = 0$ then $[AX \& S(i)]$ is a theorem of the theory, since the theory is minimally adequate; that is, the theory proves AX, which in turn proves $S(i)$. If $\phi_i(i) = 1$ then $[AX\&S(i)]$ is logically false, as discussed above. Therefore any algorithm to recursively separate the theorems of the theory from the logically false sentences would contradict Exercises 2.4.13 and 3.1.9 (the recursive inseparability of S_0 and S_1) since given i we can obviously go effectively to the sentence $[AX \& S(i)]$ simply by substituting the term **i** for the free variable x in the formula $[AX \& S(x)]$. (See Exercise 4.3.10.)   □

Note that if any given theory is inconsistent (including any given minimally adequate theory) then *all* sentences are provable in such a theory (using proofs by contradiction, of course). Thus for inconsistent theories the set of theorems of the theory includes the set of logically false sentences and the General Undecidability Theorem is trivially true and very uninteresting. On the other hand, consistent, minimally adequate theories provide the interesting cases for the General Undecidability Theorem. All theories, of course, prove all logically valid sentences, and thus consistent theories do not prove any logically false sentences (see Exercise 4.1.2). Therefore, for consistent, minimally adequate theories the set of theorems of the theory and the set of logically false sentences are disjoint and so the General Undecidability Theorem says that for consistent, minimally adequate theories these two sets are recursively inseparable.

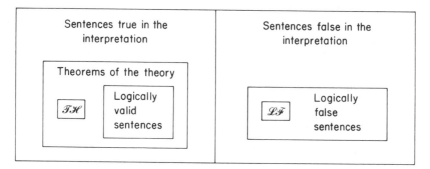

FIGURE 4.3.2 Pictorial representation of the General Undecidability Theorem.

Now let

$$\mathcal{TH} = \{AX \ \& \ S(\mathbf{i}): \phi_i(i) = 0\}$$

and

$$\mathcal{LF} = \{AX \ \& \ S(\mathbf{i}): \phi_i(i) = 1\};$$

by the previous proof these two sets are recursively inseparable. *If* we have a consistent, minimally adequate theory, and *if* we have any interpretation in which all of the axioms of that theory are true, *then* Figure 4.3.2 illustrates the relationships among some sets of sentences in a way that may be helpful in understanding the General Undecidability Theorem and its consequences. For example, if we consider the theory given by our axioms in Definition 4.1.4 (that is, given by AX) and we consider the usual interpretation consisting of the natural numbers under addition and multiplication, then Figure 4.3.2 illustrates the relationship among the set of theorems of our theory, the set of true sentences of arithmetic, the set of false sentences of arithmetic, etc.

The Gödel Undecidability Theorem for Theorems is an immediate corollary of the General Undecidability Theorem:

4.3.3 GÖDEL UNDECIDABILITY THEOREM (for Theorems) *For any consistent, minimally adequate theory there is no algorithm for deciding whether or not arbitrary sentences are theorems of the theory.*

In any reasonable mathematical theory (and for any reasonable formalization of rules of inference), the set of axioms is recursive and there are finitely many effective rules of inference. Therefore, in any *halfway reasonable* mathematical theory the sets of axioms and rules of inference will certainly both be recursively enumerable. Even under these very weak conditions, the set of theorems for such a theory will be recursively enumerable: by effectively generating a list of axioms and a list of rules of inference, one can systematically generate all possible proofs within the theory and thus list all of the theorems of the theory.

4.3.4 GÖDEL INCOMPLETENESS THEOREM   *For any consistent, minimally adequate theory such that the set of theorems of the theory is recursively enumerable, there is a sentence S such that neither S nor ~S is a theorem of the theory.*

Proof   This theorem is really a simple corollary of the Gödel Undecidability Theorem for Theorems. If there were no such sentence S then the theory would be decidable as follows: to decide whether or not a given sentence T is a theorem of the theory one simply lists the set of theorems of the theory until either T or $\sim T$ appears on the list; since the theory is consistent, exactly one of them eventually appears on the list. If T appears, it is a theorem; if $\sim T$ appears, T cannot be a theorem. Therefore, there must be such a sentence S.   □

Next we prove the Gödel Undecidability Theorem for Truth, and then finally, Church's Theorem.

4.3.5 GÖDEL UNDECIDABILITY THEOREM (for Truth)   *There is no algorithm for deciding whether or not arbitrary sentences (of the language ℒ for arithmetic) are true in the usual interpretation of the natural numbers under addition and multiplication.*

Proof   Since AX is true in the usual interpretation, the sentences in $\mathcal{TH}$ are also true in the usual interpretation, and therefore any algorithm for deciding truth in the usual interpretation would contradict the General Undecidability Theorem.   □

4.3.6 CHURCH'S THEOREM   *There is no algorithm for deciding whether or not arbitrary sentences (of ℒ) are logically valid.*

Proof   Any algorithm for deciding logical validity can be trivially converted into one for deciding logical falsity simply by negating all sentences before giving them as inputs (see Exercise 4.1.2). Since the sentences in $\mathcal{TH}$ are true in the usual interpretation and they are thus not logically false, any algorithm for deciding logical falsity would contradict the General Undecidability Theorem.   □

Additional Exercises

4.3.7   Give an algorithm for recursively separating the set of logically true sentences from the set of logically false sentences; that is, give an algorithm which given any sentence as input eventually halts with output either "true" or "false" so that all logically true sentences yield output "true" and all logically false sentences yield output "false."

4.3.8

(a) Assuming that there is at least one consistent, minimally adequate theory, show that there is a consistent, minimally adequate theory which proves a sentence which is false in the usual interpretation. *Hint*: use the Gödel Incompleteness Theorem.

(b) Assuming that there is at least one consistent, minimally adequate theory, show that there are *infinitely* many distinct (that is, pairwise contradictory) consistent, minimally adequate theories.

4.3.9   Prove a strengthened version of the Gödel Undecidability Theorem for Truth by showing that the set of sentences of $\mathscr{L}$ true in the usual interpretation is not even recursively enumerable.

*4.3.10   Using one of the programming systems of Chapter 1, explain how to write two programs such that given the integer i (in some appropriate alphabet), one program produces the sentence $[AX \, \& \, S(i)]$ of Theorem 4.3.1, while the other program produces the equivalent sentence $\exists \, x[x = i \, \& \, AX \, \& \, S(x)]$.

4.3.11   Recall from Exercise 3.1.9 that a programming system need not have a universal function in order for S_0 and S_1 to be recursively inseparable. Outline a proof of Theorem 4.3.1 which does not use the existence of a universal function, ϕ_{univ}.

Chapter 5

General Computational Complexity Theory

In this chapter we study the difficulty of computations from a very general and abstract point of view. We introduce a very general notion of a measure of computational complexity in terms of the amount of some computational "resource" a program uses in a particular computation. This notion of a general (sometimes called "abstract") computational complexity measure enables us to develop a theory of computational complexity which is insensitive to the properties of particular computing devices, programming languages, coventions for representing inputs and outputs, etc. This "machine-independent" approach lends both elegance and generality to the theory but unfortunately sacrifices the possibility of obtaining results which distinguish among different computing devices, programming languages, types of computational resources (such as time and space), etc. In Chapters 6 and 7 we deal with some less abstract versions of computational complexity theory which do make such distinctions.

Although the theory of general computational complexity measures has been highly developed and includes many interesting and important results, in this chapter we restrict ourselves to developing just a few fundamental results which both give the flavor of what has been accomplished in abstract computational complexity theory and also play a very important role in our understanding of the theory of algorithms. One principal purpose of this chapter is to raise the question: "Given a partial recursive function, what is its computational complexity?" One's response to this question is likely to be: "Given a particular general complexity measure, that is given a particular choice of computational resource, the computational complexity of a partial recursive function is given by that program for computing that function which uses the smallest amount of the specified computational resource; that is, the complexity of a partial recursive function is given by the 'best' program for computing that function." In fact, for arbitrary partial recursive functions this answer is naive and misleading: one of the main results of this chapter is the Blum Speedup Theorem, which asserts that for any complexity measure and any standard of "best use of resources," there are total recursive functions which have no best program for computing them. Roughly speaking, there are some computable functions which are so complicated that no matter which program one chooses for computing them, there is another program for computing them which is much more efficient in its use of these resources, and another which is much more efficient than that one, and another . . . *ad infinitum*.

In the last section of this chapter we prove some results which indicate the "futility of optimization" when dealing with acceptable programming systems. Of course, when the term "optimization" is used in computer science it is generally not intended to be interpreted in the literal sense of, for example, always producing the absolutely best code; rather, some very much weaker sense is intended. The results we prove begin to show just how weak this sense must be when one is dealing with arbitrary programs in an acceptable programming system.

These results we have been discussing may initially seem to be mathematically interesting but essentially "negativistic" in that they indicate how bad things are and what one cannot hope to accomplish, but with a little reflection they can be seen to be both positive and useful. Just as with unsolvability results, these results point the way to, and underscore the necessity for, developing approaches to programming which overcome these limitations; such developments are among the major goals of computer science.

Before we begin our study of general computational complexity measures, we use the first section to consider briefly an example of a particular measure of computational complexity based on our RAM programming system; one reason for doing this is to introduce in a less abstract setting some of the ideas used later in the chapter.

5.1 RAM PSEUDOSPACE MEASURE AND SIZE OF PROGRAMS

In this section we introduce and study a specific, fairly natural computational complexity measure based on our RAM programming system from Section 2.4. Our use of the term "natural" is not to be taken too literally, but rather as meaning that we are considering a computational resource for which it is *fairly reasonable* to measure the amount of the resource used in a computation.

Let $\mathbf{P}_i$ be the ith RAM program from our indexing of RAM programs in Section 2.4. We define the *pseudospace* function S_i for $\mathbf{P}_i$ (somewhat arbitrarily) as follows: if $\phi_i(x)$ is convergent then $S_i(x)$ is the largest integer in any of the RAM registers during $\mathbf{P}_i$'s computation of $\phi_i(x)$; if $\phi_i(x)$ is divergent then $S_i(x)$ is divergent. Note that Exercise 2.4.10 shows that each space function S_i is a partial recursive function. We use the convenient notational convention of sometimes letting ∞ stand for the "value" of a divergent computation, and writing $n \leq \infty$ for all natural numbers n (and also for $n = \infty$). Note that directly from this definition we have that $S_i(x)$ is convergent if and only if $\phi_i(x)$ is convergent. Also notice that if $\phi_i(x)$ is divergent then $S_i(x) = \infty$ which is not necessarily the actual amount of space used by P_i on input x; an infinite loop need not generate arbitrarily large numbers. However, any "pricing" structure certainly *should* charge infinitely much for an infinitely long computation.

What we have in mind with this approach to computational complexity is not the complexity of a program *per se* (that is, the intricacy of a program's "structure"), but rather a program's *use of a computational resource* such as time, memory space, money, etc. Since a program's use of resources depends on its actual execution on a *particular input*, the complexity of a program in this sense must be a function of the input to the program as well as of the program. This is why we have a pseudospace *function* S_i associated with each program P_i.

You may well believe that if $\phi_i(x)$ is convergent then $S_i(x)$ does not really represent the actual space used by program P_i in computing $\phi_i(x)$; the actual space used should be the maximum sum of all integers in the RAM registers at any point in P_i's computation as in our definition of RAM*space* in Section 1.9. (Recall that we are using our RAM programming system from Section 2.4 which uses RAM programs over a one letter alphabet and integers are represented in unary notation. RAM programs over larger alphabets would decrease space usage by a logarithmic factor.) However, for the purposes of this section, we reject such objections. One reason is that this more appropriate measure of space could differ by at most a multiplicative constant from S_i, where the constant is simply the numbers of registers named in P_i; but the major reason is that one goal of this chapter is to study *general* computational complexity measures in a way which ignores such minor and somewhat arbitrary details. We are studying this RAM pseudospace measure as an illustrative example of a general computational complexity measure, not because we are claiming that it is a *specific* measure of computational complexity of any particular importance, practical or otherwise. Our results on general complexity measures later in this chapter apply to *all* reasonable measures of the use of computational resources, including our RAM pseudospace measure. For this reason we are sometimes quite sketchy in our discussion of the RAM pseudospace measure in this section, particularly in the proofs of propositions which follow as special cases of later results or which are not used later in the book.

5.1.1 PROPOSITION $S_i(x) \leq y$ *is a primitive recursive predicate of* i, x, *and* y.

Proof   Let P_i use k registers and have m instructions. Then there are

$(y+1)^k$ configurations of the registers used by $\mathbf{P}_i$ in which their contents do not exceed y. Therefore, if $\mathbf{P}_i$ runs for more than $m \cdot (y+1)^k$ steps without using integers greater than y, then it must be in an infinite loop, since it has repeated a memory configuration while at the same location in the program. Thus to find out whether $S_i(x) \leq y$ one "merely" runs $\mathbf{P}_i$ on input x for at least $m \cdot (y+1)^k$ steps, watching the size of the integers in the registers which $\mathbf{P}_i$ uses. Recall that by the coding defined in Section 2.4 and by the definition of pairing functions it follows that both k and m are less than or equal to i. Therefore, formally

$$S_i(x) \leq y \quad \text{iff} \quad \forall z \leq i \forall w \leq i \cdot (y+1)^i [\Pi(z, i, \Pi_2(Comp(i, \langle x, 0 \rangle, w))) \leq y]$$
$$and \quad \exists w \leq i \cdot (y+1)^i [\Pi_1(Comp(i, \langle x, 0 \rangle, w)) = Ln(i)].$$

5.1.2 EXERCISE Referring back to Section 2.4, and in particular to Proposition 2.4.3 and its proof, verify that the formal definition of $S_i(x) \leq y$ given above is correct. □

Note that with our definition of the RAM pseudospace functions, no computation can use less space than its output; that is, $\phi_i(x) \leq S_i(x)$. Thus if some program uses no more space than its output, such a program is "optimal" in the sense that no other program can perform the same computations more "economically" in terms of our pseudo-space measure. The next proposition shows that each *pseudospace function S_i* has such an optimal program. One of the main goals of this chapter is to investigate the extent of this phenomenon: which computable functions have best programs, and in what sense are these programs best? The Blum Speedup Theorem later in this chapter establishes the existence of computable functions which have no best program in any reasonable sense.

5.1.3 PROPOSITION *For each RAM program $\mathbf{P}_i$ there is a program $\mathbf{P}_j$ such that $\phi_j = S_j = S_i$.*

Proof $\mathbf{P}_j$ is gotten from $\mathbf{P}_i$ by adding a "monitor" register which keeps track of the maximum integer which has been used by $\mathbf{P}_i$ during the course of its computation; the most obvious way to do this is to add subroutines to $\mathbf{P}_i$ after each **add** Rm instruction which compare the contents of Rm to that of the monitor register, and if necessary add 1 to the monitor. When $\mathbf{P}_i$ *would* halt, $\mathbf{P}_j$ then copies the contents of the monitor into R1 (the output register) and then halts itself. Note that j can in fact be gotten as a primitive recursive function of i.

5.1.4 EXERCISE Fill in the missing details from the proof of the previous proposition, including showing how to obtain j as a primitive recursive function of i; compare the work this involves with your (presumably) simpler proof of Exercise 2.4.10, which makes a considerably weaker assertion. ☐

The previous proposition shows that in this measure, the difficulty of computing our pseudospace functions is precisely their value. Since there are very "large" total recursive functions ϕ_i, and since $\phi_i(x) \leq S_i(x)$ for all x, it follows that there are total recursive functions which are intrinsically difficult to compute in the sense that the space used by *any* program which computes them must be very "large"; but this is hardly surprising since it is obviously very difficult to output the values of a very "large" function. This raises the question of whether there are computable functions which are "intrinsically" much more difficult to compute than the size of their values. Any programmer can write a program which uses a great deal of space (or time, or money) and yields a very small output; however, the question is whether all equivalent programs must have this same type of "expensive" behavior. The next proposition shows that there are computable characteristic functions which are arbitrarily difficult to compute in the following sense: given any total recursive function t, no matter how large, there is a computable characteristic function such that any program which computes the characteristic function must use more than $t(x)$ space for infinitely many input values x.

.5.1.5 PROPOSITION *For every total recursive function t there is a total recursive characteristic function f such that if $\phi_i = f$ then $S_i(x) > t(x)$ for infinitely many values of x.*

Proof The idea of the proof is simply to make f different from all the functions computed by programs which "generally" use less than $t(x)$ space; specifically, define

$$f(x) = \begin{cases} 1 & \text{if } S_{\Pi_1(x)}(x) \leq t(x) \text{ and } \phi_{\Pi_1(x)}(x) \neq 1 \\ 0 & \text{otherwise.} \end{cases}$$

By Proposition 5.1.1 and the fact that $\phi_{\Pi_1(x)}(x)$ is convergent if and only if $S_{\Pi_1(x)}(x)$ is convergent, f is a total recursive function. Suppose that $\mathbf{P}_i$ is any program which computes f; that is, suppose that $\phi_i = f$. Then for each of the infinitely many x's such that $\Pi_1(x) = i$ the definition of f guarantees that $t(x) < S_i(x)$. ☐

5.1.6 EXERCISE Use the previous proposition to show that there is no total recursive function g such that for all i and x, if $\phi_i(x)$ is convergent then $S_i(x) \leq g(x, \phi_i(x))$.

As we have said, one of the main results of this chapter is the Blum Speedup Theorem which shows that for every general computational complexity measure (including our RAM pseudospace measure) there are computable functions f with the property that no matter what program is used for computing f there is always another program for computing f which is much more efficient. Our next proposition shows that in our RAM pseudospace measure *every* "moderately" complex program can be replaced (effectively) by a somewhat improved equivalent program which uses only the square root of the amount of space the original program uses; and of course, the improved program could then be replaced by yet another program which is even better. This phenomenon is sometimes known as "global" speedup. We have observed that by the definition of our RAM pseudospace measure, $\phi_i(x) \leq S_i(x)$ for all i and x; on the input side, the definition also yields immediately that $x \leq S_i(x)$ for all i and x; thus $max\{x, \phi_i(x)\} \leq S_i(x)$ for all i and x, and this "input/output" restriction ultimately puts an end to any possible reduction in the amount of space a program might use to compute $\phi_i(x)$.

5.1.7 PROPOSITION *For all i we can (primitive recursively) find a j such that $\phi_i = \phi_j$ and such that for all x*

$$S_j(x) \leq max\{x, \phi_i(x), \sqrt{S_i(x)}\}.$$

That is, subject to input/output limitations, every program can be improved to use the square root of the space it used to use.

Proof The key idea of the proof is to recall that $(n+1)^2 = n^2+2n+1$, and therefore every integer w can be written uniquely as $w = x^2+2y+z$ where $x = \sqrt{w}$, $y = (w-x^2)/2$, $z = w-(x^2+2y)$. (Here as explained in the discussion between Exercises 2.13 and 2.14, we regard $\sqrt{w}$ as the greatest integer less than or equal to the (positive) real square root of w and $(w-x)/2$ as the greatest integer less than or equal to the rational $w-x$ divided by 2.) In these equations, x^2, y^2, and z^2 are all less than or equal to w. Basically, the improved program $\mathbf{P}_j$ is gotten from $\mathbf{P}_i$ by replacing each register R in $\mathbf{P}_i$ by three registers R_x, R_y, and R_z in $\mathbf{P}_j$; $\mathbf{P}_j$ then simulates the computation of $\mathbf{P}_i$ in such a way that when register R in $\mathbf{P}_i$ contains w, then the registers R_x, R_y, and R_z in $\mathbf{P}_j$ contain x, y, and z as above, respectively. Then with the obvious input/output restrictions, $\mathbf{P}_j$ will always use integers less than or equal to the square root of those used by $\mathbf{P}_i$.

5.1.8 EXERCISE Supply as many of the details which are missing from the proof of the previous proposition as are necessary to convince yourself that it is correct. □

Notice that the program $\mathbf{P}_j$ in the previous proposition would still be significantly more "efficient" even under some more "realistic" measures, such as *total space* defined as the largest integer used times the number of registers used, or *total space-time* defined as the number of instruction executions times total space.

5.1.9 EXERCISE   Verify, in detail, the claim we have just made; that is, calculate as precisely as you can the amount of improvement in P_j over P_i under both the total space and the total space-time measures.

It is natural to ask whether there is some significantly better global speedup for our RAM pseudospace measure than that given by Proposition 5.1.7. The answer is "no," and we can begin to see why by carefully examining the proofs of Propositions 5.1.1 and 5.1.5 in the light of Proposition 5.1.3. Suppose that the function t in Proposition 5.1.5 is itself a space function S_j, and that our pairing function is defined by $\langle x, y \rangle = 2^x(2y + 1) - 1$. Then any program which computes f must use at least $S_j(x)$ space for infinitely many x's; we claim that there is a "natural" program which computes the function f of Proposition 5.1.5 in "not much more" space than $S_j(x)$ for all x. If you grant us our claim, then certainly that specific program cannot be replaced by an equivalent one which uses "much less" space for all x. Specifically, our natural program for computing f simply follows the natural algorithm implied by the definition of f: first it computes $t(x)$, and since t is a space function S_j this can be done in space $S_j(x)$ by Proposition 5.1.3; then it simulates $P_{\Pi_1(x)}$ on input x to see whether $S_{\Pi_1(x)}(x) \leq S_j(x)$ and whether $\phi_{\Pi_1(x)}(x) = 1$. The second answer is a natural byproduct of obtaining the first, and from the proof of Proposition 5.1.1 we see that if $P_{\Pi_1(x)}$ uses k registers and has m instructions, then the simulation might need to be run for as many as $m(S_j(x) + 1)^k$ steps to obtain the first answer. By examining our pairing function and the way we coded RAM programs it can be shown that $2^{2^{2^k}} < x$ and $2^{2^m} < x$, and so it is sufficient to run the simulation for

$$log_2 log_2 x(S_j(x) + 1)^{log_2 log_2 log_2 x} < (S_j(x) + 1)^{(log_2 log_2 log_2 x) + 1}$$

steps, but counting this high is not our major problem. *Our* program can only have a *fixed* number of registers, and k, the number of registers in the program it is simulating can be *arbitrarily large*. So we need some way of "packing" the contents of k registers into one in a fairly efficient way; this can be done, for example, by using the pairing function from Exercise 2.1.10 so that k numbers each less than n will result in a number less than n^{2^k} in the register. Thus we can keep track of the contents of all k registers in $P_{\Pi_1(x)}$ by using numbers no larger than $(S_j(x) + 1)^{log_2 log_2 x}$. With careful programming and calculating, the sketch we have just given can be turned into a proof of the following proposition.

5.1.10 PROPOSITION *For every total recursive space function S_j there is a total recursive characteristic function f such that if $\phi_i = f$ then $S_i(x) > S_j(x)$ for infinitely many x's, and such that there is a program P_k such that $\phi_k = f$ and for all x*

$$S_k(x) \le [S_j(x) + 1]^{log_2log_2x + 1}.$$

The previous proposition, and more strongly its improvement given in Exercise 5.1.13, show that Proposition 5.1.7 yields the maximum amount of global speedup possible for our RAM pseudospace measure.

We conclude this section by considering very briefly the "size" of RAM programs. Define the *size* of a RAM program to be the total number of occurrences of characters in it.

5.1.11 PROPOSITION *Let f be a function such that if $f(n) = i$ then P_i is a shortest program such that $\phi_i(0) = n$. Then f is not recursive. That is, there is no algorithm for finding a shortest program which "prints n."*

5.1.12 EXERCISE Prove the previous proposition by modifying and formalizing the argument in the Introduction of this book.

Additional Exercises

5.1.13 Let g be any "easily" computable unbounded function from N onto N. Show that no matter what pairing function is used to obtain the coding of RAM programs, for every total recursive space function S_j there is a total recursive characteristic function f such that if $\phi_i = f$ then $S_i(x) > S_j(x)$ for infinitely many x's, and such that there is a program P_k such that $\phi_k = f$ and for all x

$$S_k(x) \le [S_j(x) + 1]^{g(x)+1}.$$

5.1.14 Show that for every total recursive function t there is a total recursive function f such that $f(x) \le x$ for all x and such that if $\phi_i = f$ then $S_i(x) > t(x)$ for *all* $x > i$. (Contrast this with Proposition 5.1.5.)

5.1.15 Use the Recursion Theorem to prove that for any total recursive function t and for any partial recursive function ψ we can find a program P_i which computes ψ (that is, $\phi_i = \psi$) and such that $S_i(x) \ge t(x)$ for all x; (it is *cheating* to solve this by simply showing how to write such "bad" programs.)

5.2 GENERAL COMPUTATIONAL COMPLEXITY MEASURES

In this section we introduce and begin to study the notion of general computational complexity measures. This general approach uses only two defining conditions and is analogous to our definition of acceptable programming systems in Section 3.1. These two conditions are quite obviously true of all "natural" measures of computational complexity (including the RAM pseudospace measure considered in the previous section), and there are many interesting and important properties of computational complexity which can be derived from them. Our main result in this section is the Recursive Relatedness Theorem for general complexity measures, and we conclude the section with some applications of that theorem. We now give the definition of computational complexity measures, which is due to Manuel Blum.

5.2.1 DEFINITION Let ϕ_0, ϕ_1, . . . be any acceptable programming system. A listing Φ_0, Φ_1, . . . of partial recursive functions is a *computational complexity measure* (on the given acceptable programming system) if it satisfies the following conditions:

1. For all i and x, $\phi_i(x)$ is convergent if and only if $\Phi_i(x)$ is convergent.
2. $\Phi_i(x) \leq y$ is a recursive predicate of i, x, and y.

Our RAM pseudospace functions S_0, S_1, . . . are a complexity measure (by Proposition 5.1.1), and by Rogers' Isomorphism Theorem we can "transfer" this measure to any other acceptable programming system. Thus every acceptable programming system has *some* computational complexity measure on it. Any other fairly reasonable notion of the use of computational resources also satisfies this definition. Some examples are the number of ALGOL instruction executions, the number of Turing machine instruction executions (TM*time*), the amount of space used for the execution of PASCAL programs, and the number of tape squares visited by a Turing machine (TM*space*). But not everything is a complexity measure: if $\Phi_i(x) = \phi_i(x)$ then Condition 1 is satisfied but Condition 2 is not (by the unsolvability of the halting problem); if $\Phi_i(x) = 0$ for all i and x then Condition 2 is satisfied but Condition 1 is not. (Incidently, these last two examples show that Conditions 1 and 2 are "independent.")

The many examples of general complexity measures make it clear that results about all complexity measures cannot enable us to distinguish among different models of computation, or among different notions of resource use on the same model, such as the differences between time and space. Even worse, one cannot distinguish the relative difficulty of computing particular functions: there are measures in which it is no more difficult to compute 2^{2^x} than to compute the zero function. In fact, let f be any total recursive function, let Φ_0, Φ_1, . . . be any complexity measure, and let i be such that $\phi_i = f$. If we define $\Psi_j(x) = \Phi_j(x)$ for all j not equal to i, and we define $\Psi_i(x) = 0$ for all x, then Ψ_0, Ψ_1, . . . is also a complexity measure and in it f has *zero* complexity. We have simply *decreed* that there is "no charge" for running program i! (Notice that once again we are completely identifying a program P_i with its integer index i.)

A straightforward extension of the argument we have just given indicates just how "unnatural" some complexity measures can be. Any sequence of partial recursive functions ψ_0, ψ_1, ... is said to be *recursively enumerable* if programs computing the functions (in an acceptable programming system ϕ_0, ϕ_1, ...) can be enumerated effectively; that is, if there is a total recursive function h such that $\psi_i = \phi_{h(i)}$ for all i.

5.2.2 PROPOSITION *Let* f_0, f_1, ... *and* g_0, g_1, ... *be any two recursively enumerable sequences of total recursive functions. For any acceptable programming system* ϕ_0, ϕ_1, ... *there is a complexity measure* Φ_0, Φ_1, ... *on that system such that for all i there is a j with* $\phi_j = f_i$ *and* $\Phi_j = g_i$.

5.2.3 EXERCISE Prove the previous proposition. *Hint*: you may want to use the results of Exercise 3.4.19.

The previous proposition shows that there are pathological complexity measures in which the complexities of (programs which compute) infinitely many given functions f_i are arbitrarily decreed to be some functions g_i, which may be ridiculous as measures of the complexities of the functions f_i. In view of the fact that such arbitrary "pricing" for computation is possible, one might question whether any useful consequences can follow from such a broad definition of complexity measure. To allay these fears somewhat we give the following useful theorem, which asserts that, nevertheless, all complexity measures are "more or less" equal (up to some computable factor r). This theorem generalizes the results of Section 1.9, which dealt only with some "natural" complexity measures. In the sense that it shows that all complexity measures are "the same" to within some recursive factor r, the Recursive Relatedness Theorem which follows is *weakly* analogous to Rogers' Isomorphism Theorem, which shows that any two acceptable programming systems are "identical" up to an effective isomorphism between the systems. Of course, unless the complexity measures we deal with are in some sense "natural," we may expect the factor r which relates the two measures to be (unreasonably) large. However, from Section 1.9 you should recall that for many natural measures a bound r satisfying the following theorem is reasonably small.

5.2.4 THEOREM (Recursive Relatedness of Complexity Measures) *Let* ϕ_0, ϕ_1, ... *and* ψ_0, ψ_1, ... *be acceptable programming systems and let t be a total recursive function translating the first to the* *second; that is, $\phi_i = \psi_{t(i)}$ for all i. Let Φ_0, Φ_1, . . . and Ψ_0, Ψ_1, . . . be complexity measures on these two acceptable programming systems, respectively. Then there is a total recursive function r such that for all i*

$$\Phi_i(x) \leq r(x, \Psi_{t(i)}(x)) \qquad and \qquad \Psi_{t(i)}(x) \leq r(x, \Phi_i(x))$$

for all but finitely many values of x. Moreover, r can be made monotone in its second argument; that is, $r(x, y) \leq r(x, y + 1)$ for all x and y.

Proof The definition of r uses a simple "max-ing" construction which is used frequently in this chapter. Define

$$h(i, x, y) = \begin{cases} max\{\Phi_i(x), \Psi_{t(i)}(x)\} & \text{if} \quad \Phi_i(x) = y \text{ or } \Psi_{t(i)}(x) = y \\ 0 & \text{otherwise.} \end{cases}$$

By Condition 2 for complexity measures and the fact that t is recursive, the predicate $[\Phi_i(x)=y$ or $\Psi_{t(i)}(x)=y]$ is recursive; by condition 1 for measures and the fact that t is a translation, $\Phi_i(x)$ is convergent if and only if $\Psi_{t(i)}(x)$ is convergent. Therefore, h is a total recursive function. For all x and y define

$$r(x, y) = max_{i \leq x} max_{z \leq y} h(i, x, z).$$

Then for each i and for all $x \geq i$ we have

$$\Phi_i(x) \leq r(x, \Psi_{t(i)}(x)) \qquad and \qquad \Psi_{t(i)}(x) \leq r(x, \Phi_i(x)),$$

and $r(x, y) \leq r(x, y + 1)$ for all x and y. $\qquad \square$

Notice that if the translation t in the previous theorem is not *onto* then, while the function r bounds the complexities of *all* programs in the first system in terms of the complexities of equivalent programs in the other system, r does not necessarily bound the complexities of *all* programs in the second system in terms of the complexities of equivalent programs in the first system. It is often convenient to have a single bounding function which works for the complexities of functions in both systems, and there are two very simple ways we could achieve this. We could switch the two systems, apply the theorem, obtain another bounding function r' which works in the other direction, and finally use $r + r'$ as a bounding function which works in both directions; but this is much too much work! All we need to do is take the translation t to be both one-to-one and onto (as supplied by Rogers' Isomorphism Theorem), and then the bounding function r in the proof works in both directions; moreover, since t^{-1} is also an isomorphism (see Exercise 3.4.14) we can effectively translate in either direction. Henceforth, unless we explicitly state otherwise, whenever we refer to the Recursive Relatedness Theorem for complexity measures we assume that the translation t is one-to-one and onto.

Since, as in the Recursive Relatedness Theorem, we shall often be asserting that some property P (or equality, or inequality, etc.) holds for all but finitely many values of x, we introduce the abbreviation $P(x)$ *a.e.* (*almost everywhere*) to mean that $P(x)$ holds for all but finitely many values of x; similarly, $P(x)$ *i.o.* (*infinitely often*) means that $P(x)$ holds for infinitely many values of x. If there are several variables and there may be some doubt as to which variable is intended, we may also write "a.e. x" or "i.o. x" to avoid any ambiguity. Note that the a.e. condition on the inequalities in the proof of the Recursive Relatedness Theorem arise from the fact that the "max-ing" construction of r "considers" more and more programs as the argument x increases so that any given program i gets considered at almost every x.

5.2.5 EXERCISE   Show that the Recursive Relatedness Theorem cannot be improved in either of the following ways:

(a) The bounding function r *cannot* be made a function of the argument y alone, yielding

$$\Phi_i(x) \le r(\Psi_{t(i)}(x)) \text{ a.e.} \qquad \text{and} \qquad \Psi_{t(i)}(x) \le r(\Phi_i(x)) \text{ a.e.}$$

(b) The inequalities on the measures *cannot* be made to hold for all values of x.

Hint: for Φ use Turing machine space and for Ψ use the number of steps a Turing machine actually executes. (How does this differ from TM*time* of Exercise 1.9.4?) For part (b), take $\Psi_i(x) = \Phi_i(x) + i$.

The previous exercise shows that the Recursive Relatedness Theorem cannot be strengthened in two potentially desirable ways. However, at least one, and sometimes both, of these improvements can often be made. For example, if we define time functions for the RAM programs of Section 2.4 by setting $T_i(x)$ equal to the total number of instruction executions during $\mathbf{P}_i$'s computation of $\phi_i(x)$, then $S_i(x) \le x + T_i(x)$ and $T_i(x) \le i \cdot (S_i(x) + 1)^i$ for *all* x. (S_i is the RAM pseudospace function from Section 5.1.)

5.2.6 EXERCISE   Verify the preceding assertion. How should the inequalities be changed if T_i is replaced by RAM*time*$_{\mathbf{P}_i}$ as defined in Section 1.9?

The Recursive Relatedness Theorem is often a useful technical tool for generalizing results from one particular complexity measure to all complexity measures; this is analogous to the use of Theorem 3.1.5 and the Rogers' Isomorphism Theorem for proving properties of acceptable programming systems. For example, the following.

5.2.7 PROPOSITION   *For any complexity measure* Φ_0, Φ_1, . . . *there is a total recursive (bounding) function b such that for all i,*

$$\phi_i(x) \le b(x, \Phi_i(x)) \text{ a.e.}$$

5.2.8 EXERCISE   Prove the previous proposition using the Recursive Relatedness Theorem and our RAM pseudospace measure.

As another example of a result that can be transferred by way of the recursive relatedness of complexity measures consider the next proposition.

5.2.9 PROPOSITION   *Let* ϕ_0, ϕ_1, . . . *and* Φ_0, Φ_1, . . . *be any acceptable programming system and any complexity measure on it. Then for any total recursive function t there is a total recursive characteristic function f such that for all i, if* $\phi_i = f$, *then* $\Phi_i(x) > t(x)$ *i.o.*

5.2.10 EXERCISE   Prove the previous proposition, using recursive relatedness and Proposition 5.1.5.

The function b in Proposition 5.2.7 bounds the output value of a computation in terms of its input and complexity; this agrees with our intuition since it certainly ought to "cost" more to "print out" larger output values. As we saw in Exercise 5.1.6, Proposition 5.2.9 shows that (for any complexity measure) there is no total recursive function g "bounding" in the reverse direction; that is, bounding the complexity of a computation in terms of its input and output values. In other words, there is no total recursive function g such that for all i,

$$\Phi_i(x) \le g(x, \phi_i(x)) \text{ a.e.}$$

Additional Exercises

*5.2.11

(a) Give the intuitive reason why the bounding function b in Proposition 5.2.7 needs to be a function of the input x as well as of the complexity of the computation, even for "natural" measures.

(b) *Prove* that Proposition 5.2.7 cannot be improved to have b be a function of the complexity $\Phi_i(x)$ alone, or to have the inequality hold for all x.

(c) Give an example of a complexity measure for which b can be taken to be a function of $\Phi_i(x)$ alone *and* such that the inequality holds for all x.

5.2.12   Using the "max-ing" type of construction in the proof of the Recursive Relatedness Theorem, prove *directly* that in any complexity measure there are total recursive functions f and g such that for all i,

$\Phi_i = \phi_{f(i)}$ and $\Phi_{f(i)}(x) \le g(x, \Phi_i(x))$ a.e. This is the general complexity measure version of Proposition 5.1.3.

5.2.13 Select several of the measures of Section 1.9. Verify that these measures have very low "overhead" for patching finite tables into programs. Specifically, show that if $\phi_i = \phi_j$ a.e., then there is a j' such that $\phi_i = \phi_{j'}$ and $\Phi_{j'}(x) \le \Phi_j(x) + 2|x|$ a.e. For which of these measures can you make $\Phi_{j'} = \Phi_j$ a.e.?

5.2.14 Note that if j is an "optimal" program for a total recursive function and $t(i) = j$ for all i, then $\phi_i = \phi_j$ implies that $t(i)$ is an "optimal" program for ϕ_i, but this hardly warrants calling t an "optimizer" for the function ϕ_j. Now show that no computable function can come much closer to being an "optimizer" for ϕ_j. Specifically, let j be any program for a total recursive function and let t be any total recursive function such that $\phi_i = \phi_j$ implies $\phi_{t(i)} = \phi_j$ and such that $\phi_{t(i)}(x)$ defined implies $\phi_i(x)$ defined. Let $r(x, y)$ be any total recursive function. Show how to find an i such that $\phi_i = \phi_j$ but $\Phi_{t(i)}(x) > r(x, \Phi_j(x))$ a.e.

Hint: Define a total recursive function f by

$$\phi_{f(i)}(x) = \begin{cases} \phi_i(x) + 1 & \text{if } \Phi_{t(i)}(x) \le r(x, \Phi_j(x)) \\ \phi_j(x) & \text{otherwise,} \end{cases}$$

and apply the recursion theorem.

5.3 GAP, COMPRESSION, AND SPEEDUP THEOREMS

In this section we begin to explore the question, "What is the complexity of a partial recursive function?" Proposition 5.1.3 shows that the RAM pseudospace functions have "optimal" programs, and it would be reasonable to say that (the complexity of) such an optimal *program* gives the complexity of a pseudospace *function*. Proposition 5.1.7 on global speedup for our RAM pseudospace measure shows that many other computable functions cannot have *absolutely* optimal programs, but Proposition 5.1.10 shows that nevertheless there are some functions which, in a fairly weak sense, do have optimal programs to within the limits allowed by global speedup; for such functions it still might be reasonable to define their complexity in terms of a (roughly) "best" program. The Compression Theorem, our second theorem of this section, is a stronger version of Proposition 5.1.10 for general complexity measures. It shows that for each measure there is a total recursive function (which might be called a "slop factor") such that there are small, arbitrarily complex total recursive functions which *do* have "best" programs to within this factor. But this phenomenon is not universal. Our final theorem of this section, the Blum Speedup Theorem, shows that in any measure and for any recursive factor (no matter how large) there are small total recursive functions which *do not* have best programs to within this factor; thus, looking for "best" programs will not provide a reasonable notion for the complexity of arbitrary partial recursive functions.

We begin this section with a very simple, but perhaps somewhat surprising result. Suppose we consider all of the functions computable within some total recursive bound t on the use of computational resources and choose some very large total recursive function g (for example, $g(x) = 2^{2^x}$). Will increasing the bound on the use of resources to $g \circ t$ (for example, $2^{2^{t(x)}}$ on input x) necessarily allow the computation of some *new functions*? The answer is "no," and in fact there may not even be any *programs* whose use of resources (that is, complexity) falls between t and $g \circ t$!

Throughout this section, unless we specifically state otherwise, ϕ_0, ϕ_1, . . . stands for any acceptable programming system and Φ_0, Φ_1, . . . stands for any general computational complexity measure on that system.

5.3.1 GAP THEOREM   *Let g be any total recursive function of two arguments such that $y < g(x, y)$ for all x and y. From (a program for) g we can effectively find (a program for) a total recursive function t such that for all i and x,*

$$\text{if } t(x) < \Phi_i(x) < g(x, t(x)) \text{ then } x \le i;$$

that is, for every program the complexities of that program can enter the "gap" between t and $g \circ t$ only finitely often.

Proof   A definition of a suitable t is quite simple to state:

$$t(x) = \min y[\forall i(y < \Phi_i(x) < g(x, y) \text{ implies } x \le i)].$$

The problem now is to show that t is a total recursive function. First note that the definition of t may obviously be rewritten as

$$t(x) = \min y[\forall i < x(\Phi_i(x) \le y \text{ or } g(x, y) \le \Phi_i(x))].$$

Given x and y, the condition in the brackets can be tested effectively by Condition 2 for complexity measures, since g is a total recursive function. Thus it is obviously sufficient to show that for each x there is *some* y which satisfies the condition in the brackets. Since $y < g(x, y)$ for all x and y, if for some given x we define $y_0 = 0$ and $y_{j+1} = g(x, y_j)$ for all j, then

$$y_j < g(x, y_j) = y_{j+1} < g(x, y_{j+1})$$

for all j. There are *at most* x (finite) values $\Phi_i(x)$ with $i < x$, and since there are $x + 1$ "g-gaps" in the sequence $y_0, \ldots, y_{x+1}$, there *must* be *some* $j \leq x + 1$ such that

$$\forall i < x(\Phi_i(x) \leq y_j \text{ or } g(x, y_j) \leq \Phi_i(x)).$$

This proves the theorem (and incidently, also shows that for each x, $t(x) \leq y_{x+1}$). ☐

5.3.2 EXERCISE Suppose that g is a primitive recursive function such that $y < g(x, y)$ for all x and y, and that $\Phi_i(x) \leq y$ is a primitive recursive predicate of i, x, and y. Show that the function t in the (proof of the) Gap Theorem is primitive recursive.

The Gap Theorem shows that for any complexity measure there are arbitrarily large gaps such that no complexity function can enter the gap more than finitely often. A rather provocative interpretation of this theorem in conjunction with the Recursive Relatedness Theorem is that given two computers, one much faster than the other, there are always time bounds t in which the slow computer can catch the fast one! Let r be a total recursive function relating the two time measures for these computers and let t be a total recursive function at the bottom of an r-gap in the time measure for the slow computer. Then every program which runs in time t on the fast computer runs in time $r \circ t$ on the slow computer. But every program which runs in time $r \circ t$ on the slow computer must really run in time t on the slow computer. Therefore every program which runs in time t on the fast computer also runs in time t on the slow computer!

It is natural to ask whether the situation the Gap Theorem presents is not somewhat pathological, particularly in light of the previous paragraph. The next theorem shows that for any complexity measure there is some fixed recursive factor g such that if t is a total *complexity* function then there will be new, fairly small functions one can compute by raising the resource bound from t to $g \circ t$. This contrasts strongly with the Gap Theorem since it shows that for each such function t, not only are there *programs* whose complexity lies between t and $g \circ t$, but also to within the factor g, these programs are *optimal* for the functions which they compute.

5.3.3 COMPRESSION THEOREM *There are total recursive functions f and g such that for all i and x, if $\Phi_i(x)$ is convergent then $\phi_{f(i)}(x)$ is* *convergent and* $\phi_{f(i)}(x) \leq x$, *and such that*

1. For all j, if $\phi_j = \phi_{f(i)}$ *then* $\Phi_j(x) \geq \Phi_i(x)$ *a.e.*

2. $\Phi_{f(i)}(x) \leq g(x, \Phi_i(x))$ *a.e.*

Proof   The object of the proof is to define $\phi_{f(i)}$ so that it differs from the result of each program which has complexity less than $\Phi_i(x)$ infinitely often. This is done using sets $C(i,x)$ of *cancelled* programs (whose complexity is too low). Define

$$C(i, x) = \{j < x : \Phi_j(x) \leq \Phi_i(x)\},$$

and define

$$\phi_{f(i)}(x) = \begin{cases} min \ y[y \neq \phi_j(x) \ for \ all \ j \in C(i, x)]. & if \ \ \phi_i(x) \ convergent \\ divergent & otherwise. \end{cases}$$

If $\Phi_i(x)$ is convergent, then Condition 2 for measures insures that we can compute $C(i, x)$ and Condition 1 insures that $\phi_j(x)$ is convergent for all j in $C(i, x)$; thus $\phi_{f(i)}(x)$ is convergent in this case, and obviously $\phi_{f(i)}(x) \leq x$. If $\Phi_i(x)$ is divergent, then $\phi_{f(i)}(x)$ will be divergent. Of course we get the total recursive function f by our standard *s-m-n* construction from Chapter 3, and part 1 follows immediately from the definition of $\phi_{f(i)}$.

We prove part 2 using a very simple "max-ing" construction of the type used to prove the Recursive Relatedness Theorem. Define

$$h(i, x, y) = \begin{cases} \Phi_{f(i)}(x) & if \ \ \Phi_i(x) = y \\ 0 & otherwise, \end{cases}$$

and let $g(x, y) = \max_{i \leq x} h(i, x, y)$. Then if $\Phi_i(x)$ is convergent and $x \geq i$,

$$\Phi_{f(i)}(x) = h(i, x, \Phi_i(x)) \leq g(x, \Phi_i(x));$$

if $\Phi_i(x)$ is divergent then

$$\Phi_{f(i)}(x) = \infty \leq \infty = g(x, \Phi_i(x))$$

by our convention. This completes the proof of the theorem. $\square$

The Compression Theorem shows that there are arbitrarily complex (fairly small) computable functions whose *complexity* is well-defined to within a factor of g in the sense that these functions have "optimal" programs for computing them to within the factor g (which depends *only* on the complexity measure).

One natural way to view the a.e. condition in the Compression Theorem is that it is unavoidable because of the possibility of including finite tables in programs in order to compute finitely many values of a function with very little use of resources (that is, by "table lookup"). Although this is a natural interpretation, we cannot prove that it is the only interpretation. One way to get sharp theoretical results about the inclusion of finite tables in programs is to have a reasonable theory of the size of programs, a project we begin in the next section.

If you worked Exercise 5.2.12 you saw that in any complexity measure there is a fixed recursive factor such that the *complexity functions* for that measure have well-defined complexities in the sense of having optimal programs to within that fixed factor. The Compression Theorem shows that in addition there are small computable functions whose complexity is similarly well-defined in terms of optimal programs. This still leaves open the question of whether there is some recursive factor such that *every* total computable function has an optimal program to within that factor. The Blum Speedup Theorem answers this question in the negative, showing that for any complexity measure there are some computable functions for which not even "near optimal" programs exist. Before we prove the full Blum Speedup Theorem we prove a slightly weaker version, the proof of which is *notationally* much simpler, and then we use this weaker version to prove the full Blum Speedup Theorem.

5.3.4 RESTRICTED SPEEDUP THEOREM *Let* Φ_0, Φ_1, . . . *be a complexity measure satisfying the condition*

(a) $$\Phi_{s(i,x)}(y) \leq \Phi_i(x, y) \text{ for all } i, x, \text{ and } y,$$

where s is an s-1-1 function for the acceptable programming system ϕ_0, ϕ_1, ... *Then for every total recursive function g such that* $g(x, y) \leq g(x, y + 1)$ *for all x and y, there is a total recursive function f such that* $f(x) \leq x$ *for all x and such that if* $\phi_i = f$ *then there is a j such that* $\phi_j(x) = f(x)$ *a.e. and*

$$g(x, \Phi_j(x)) \leq \Phi_i(x) \text{ a.e.}$$

Before giving the proof of this theorem we would like to point out that g may be a very "large" function. For example, if $g(x, y) = (x \cdot y)^{x \cdot y}$ then the theorem asserts that if i is any program for computing f then there is another program j which *nearly* computes f (and perhaps could be "patched" with a finite table so that it does compute f; see Exercise 5.2.13) such that $(x \cdot \Phi_j(x))^{x \cdot \Phi_j(x)} \leq \Phi_i(x)$ a.e. Of course, one would rather use program j (appropriately patched) for computing f rather than program i. But then notice that the theorem could then be applied again to show that the patched version of program j is still far from optimal.

Notice also that our RAM pseudospace measure and all of the space measures of Section 1.9 satisfy condition (a) in the Restricted Speedup Theorem. In fact, all "natural" measures, including all of the time measures of Section 1.9, come close to satisfying condition (a) in that there is a *small* recursive function h such that for all i, x, and y

$$\Phi_{s(i,x)}(y) \leq \Phi_i(x, y) + h(x, y);$$

this is because all the function s does is add a preprocessor to the beginning of program i which when it gets input y computes the coded pair $\langle x,y \rangle$ and then gives this as input to program i. (We explore this more fully in Chapter 6.)

5.3.5 EXERCISE Show that our RAM pseudospace measure, RAM-*space*, and TM*space* all satisfy condition (a) in the Restricted Speedup Theorem. If you find this exercise too difficult, see its restatement at the end of this section for a hint.

Proof To prove the Restricted Speedup Theorem, we shall define a program n computing a partial recursive function ϕ_n which we shall think of as having two arguments, and we shall define the function f by $f(x) = \phi_n(0, x)$ for all x. $\phi_n(w, x)$ will be defined recursively from $\phi_n(w, 0), \ldots, \phi_n(w, x-1)$, and if $w < x$, also recursively from $\phi_n(x, x)$, $\phi_n(x-1, x), \ldots, \phi_n(w+1, x)$. Thus, since our program n is "introspective," we shall use the Recursion Theorem to justify this simple type of recursive definition in our arbitrary acceptable programming system.

The intuitive discussion in this paragraph is intended to be *helpful*; if you find it confusing, ignore it (for the time being) and continue with the proof of the theorem. The idea behind the proof is as follows: for $w < x$ the program n "tries" *both* to make $\phi_n(w, x) = \phi_n(w+1, x)$ and also to *cancel w if* $\Phi_w(x) < g(x, \Phi_n(w+1, x))$. If $\Phi_w(x) < g(x, \Phi_n(w+1, x))$ then program n cancels w by guaranteeing that $\phi_n(w, x) \neq \phi_w(x)$. It is this step which guarantees speedup, but because of it the program may not be able to make $\phi_n(w, x) = \phi_n(w+1, x)$. However, once w has been cancelled, n no longer "worries" about its complexity, and so, as functions of the argument x, $\phi_n(w, x)$ and $\phi_n(w+1, x)$ are almost the same function. Furthermore, if the program w is not *much* "slower" than $\Phi_n(w+1, x)$ (by the factor g) on input x, then the cancellation of w guarantees that program w does *not* compute $\phi_n(w, x)$ (considered as a function of x). Now here is the proof:

To begin, for *any* n, w, and x we define the set of programs *cancelled at stage x by n and w*, $C(n,w,x)$ to be

$$\{i: w \leq i < x, i \notin \bigcup_{y<x} C(n, w, y), \text{ and } \Phi_i(x) < g(x, \Phi_n(i+1, x))\}.$$

It is our explicit intention that the conditions in the definition of $C(n, w, x)$ be tested "sequentially," thus if $x \leq w$ then $C(n, w, x) = \emptyset$ for all n. Therefore, given fixed n and w, if $C(n,w,y)$ is defined for all $y < x$

and $\Phi_n(i+1, x)$ is defined for all i such that $w \leq i < x$ then $C(n,w,x)$ is defined and its members can be determined effectively.

Next, we define (using our standard s-m-n construction) a total recursive function h such that for all n, w, and x

$$\phi_{h(n)}(w, x) = min\ y[\ y \neq \phi_i(x)\ for\ all\ i \in C(n, w, x)].$$

Notice that if $\phi_{h(n)}(w, x)$ is defined then $\phi_{h(n)}(w, x) \leq x$ since there are at most x elements in $C(n,w,x)$. By the Recursion Theorem, now let n be a fixed program such that $\phi_n = \phi_{h(n)}$. Figure 5.3.6 gives a flow chart for computing ϕ_n. From the previous paragraph we have that $\phi_n(w,x) = 0$ whenever $x \leq w$. Now let us see what is required to have $\phi_n(w,x)$ defined when $x > w$. First $C(n,w,x)$ must be defined. For this we need to know that $C(n,w,y)$ is defined for all $y < x$ and that $\Phi_n(i+1,x)$ is defined

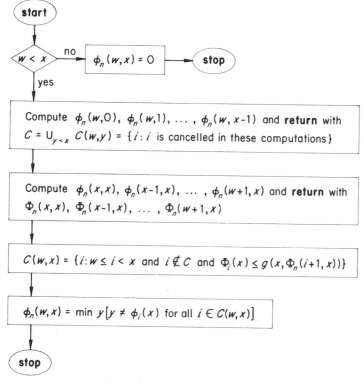

FIGURE 5.3.6   A flow chart for computing $\phi_n(w, x)$. (Since n is fixed via the Recursion Theorem, $C(n, w, x)$ is abbreviated to $C(w, x)$.)

for certain i such that $w \leq i < x$. The latter is guaranteed if $\phi_n(x,x)$, $\phi_n(x-1,x)$, ..., $\phi_n(w+1,x)$ are all defined. Because of the definition of $\phi_{h(n)}(w,y)$, the former is guaranteed if $\phi_n(w,0)$, ..., $\phi_n(w,x-1)$ are all defined. Finally $\phi_i(x)$ *must be defined for all* $i \in C(n,w,x)$. This is guaranteed by the definition of $C(n,w,x)$ and the fact that $C(n,w,x)$ has already been guaranteed to be defined.

Since $\phi_n(w, 0) = 0$ for all w, we now proceed by induction on x to show that $\phi_n(w, x)$ is defined for all w and x. Assume that $x > 0$ and that $\phi_n(w, y)$ is defined for all w and all $y < x$. Recall that $\phi_n(w, x) = 0$ for all $w \geq x$, and so by the criterion in the previous paragraph $\phi_n(x-1, x)$ must *also* be defined. By the same reasoning $\phi_n(x-2, x)$ is defined, and we may continue *downward* until we get that $\phi_n(0,x)$ is defined. Therefore, ϕ_n is a total recursive function.

By inspecting the definition of $C(n,w,x)$ it is very easy to see by induction on x that for all $w > 0$,

$$C(n, w, x) = C(n, 0, x) - \{0, \ldots, w - 1\}.$$

Since each program i can be in at most one of the sets $C(n, 0, 0)$, $C(n, 0, 1)$, $C(n, 0, 2)$, ... (once a program is cancelled, it never can be cancelled again), for each w there is a number m_w such that, if $i < w$ and $i \in \cup_{y < \infty} C(n, 0, y)$, then $i \in \cup_{y < m_w} C(n, 0, y)$; that is, every $i < w$ that is ever cancelled by n and 0 is cancelled at some stage less than m_w. Therefore, if $x \geq m_w$ we have $C(n, 0, x) = C(n, w, x)$; and so by the definition of ϕ_n, for all $x \geq m_w$, $\phi_n(0, x) = \phi_n(w, x)$. Note that we are *not* claiming that m_w can be determined effectively from w (in fact, it *cannot*), but merely that there is an m_w which works.

Finally, define $f(x) = \phi_n(0, x)$. Suppose for some i and for some $x > i$ that $g(x, \Phi_n(i + 1, x)) > \Phi_i(x)$; then in the calculation of f we *cancel* i for the least such $x > i$ (because $i \in C(n, 0, x)$ for that x) yielding $f(x) = \phi_n(0, x) \neq \phi_i(x)$. Therefore, we have that if $\phi_i = f$, then

$$g(x, \Phi_n(i+1, x)) \leq \Phi_i(x) \qquad \text{for all } x > i.$$

From this, from condition (a), and from the fact that g is nondecreasing in its second argument, we have that if $\phi_i = f$ then

(*) $$g(x, \Phi_{s(n,i+1)}(x)) \leq g(x, \Phi_n(i+1, x)) \leq \Phi_i(x) \text{ a.e.}$$

and

$$\phi_{s(n,i+1)}(x) = \phi_n(i+1, x) = \phi_n(0, x) = f(x) \text{ a.e.}$$

Letting $j = s(n, i+1)$ completes the proof of the theorem. $\square$

Before we proceed to the full Blum Speedup Theorem, we need an indexing $\psi_0, \psi_1, \ldots$ of the finite functions whose domains are the nonempty initial segments of the natural numbers; one such suitable indexing was given in Exercise 2.1.12a. In any acceptable programming system ϕ_0, ϕ_1, . . . there is a total recursive function k such that

$$\phi_{k(i,y)}(x) = \begin{cases} \psi_y(x) & \text{if} \quad x \in \text{domain } \psi_y \\ \phi_i(x) & \text{if} \quad x \notin \text{domain } \psi_y \end{cases}$$

for all i, x, and y. In any "natural" complexity measure, for any reasonable such function k there is a *small* recursive function h' such that for all i, x, and y

$$\text{if} \quad x \notin \text{domain } \psi_y \quad \text{then} \quad \Phi_{k(i,y)}(x) \le \Phi_i(x) + h'(x, y);$$

this is because such a function k merely adds a preprocessor to the beginning of program i which, when it gets input x, first checks to see if $x \in \text{domain } \psi_y$ and then either does a table lookup if it is, or else turns x over as input to program i if it is not. In fact, all of our space measures satisfy the following even stronger condition:

(b) There is a function k such that for all i, x, and y

$$\phi_{k(i,y)}(x) = \begin{cases} \psi_y(x) & \text{if} \quad x \in \text{domain } \psi_y \\ \phi_i(x) & \text{if} \quad x \notin \text{domain } \psi_y \end{cases}$$

and

$$\text{if} \quad x \notin \text{domain } \psi_y \quad \text{then} \quad \Phi_{k(i,y)}(x) \le \Phi_i(x).$$

When we prove the Blum Speedup Theorem below we shall first prove it for complexity measures satisfying conditions (a) and (b), where condition (a) comes from the statement of the Restricted Speedup Theorem.

5.3.7 EXERCISE Show that for our RAM pseudospace measure there is a primitive recursive function k satisfying condition (b).

5.3.8 BLUM SPEEDUP THEOREM *For any complexity measure and any total recursive function g such that $g(x, y) \le g(x, y + 1)$ for all x and y the following holds: there is a total recursive function f such that $f(x) \le x$ for all x and such that if $\phi_i = f$ then there is a j such that $\phi_j = f$ and*

$$\left[g(x, \Phi_j(x)) \le \Phi_i(x) \text{ a.e.} \right]$$

Proof First, we assume that the complexity measure satisfies conditions (a) and (b); (by Exercises 5.3.5 and 5.3.7, our RAM pseudospace measure is one such measure). Continuing from the end of the proof of the Restricted Speedup Theorem, we note that since for all w, $\phi_n(0, x) = \phi_n(w, x)$ a.e., then for all w there is some v such that $f = \phi_{k(s(n,w),v)}$; that is, there is *some* way to "patch" the program $s(n,w)$ with a "finite table'' so that the resulting program computes f. Then by conditions (a) and (b) and by the inequality (*) from near the end of the proof of the Restricted Speedup Theorem, and by the fact that g is nondecreasing in its second argument, we have that if $\phi_i = f$ then for some v,

$$g(x, \Phi_{k(s(n,i+1),v)}(x)) \le g(x, \Phi_{s(n,i+1)}(x)) \le \Phi_i(x) \text{ a.e.}$$

Letting $j = k(s(n, i + 1), v)$ completes the proof for measures satisfying conditions (a) and (b).

One way to prove this theorem for all complexity measures is to replace $\Phi_n(i+1,x)$ by $\max_{v \le x} \Phi_{k(s(n,i+1),v)}(x)$ in the definition of $C(n,w,x)$ and verify that essentially the same proof then works. Another way is to use the recursive relatedness of measures. We use the latter approach.

Let g be the amount of speedup we wish in the complexity measure $\Phi_0, \Phi_1, \ldots$ on the acceptable programming system $\phi_0, \phi_1, \ldots$. Let t be a total, one-to-one, and onto translation from $\phi_0, \phi_1, \ldots$ to our RAM programming system $\theta_0, \theta_1, \ldots$. Let $S_0, S_1, \ldots$ be our RAM pseudospace measure (for which we now know that the Blum Speedup Theorem holds), and let r be a total recursive function given by the Recursive Relatedness Theorem such that for all i,

$$\Phi_i(x) \le r(x, S_{t(i)}(x)) \quad \text{and} \quad S_{t(i)}(x) \le r(x, \Phi_i(x)) \quad \text{a.e.,}$$

and such that for all x and y, $r(x, y) \le r(x, y + 1)$. Define

$$g'(x, y) = r(x, g(x, r(x, y)))$$

and take f to have g' speedup in our RAM pseudospace measure; that is $f(x) \le x$ for all x and if $\theta_{i'} = f$ then there is a j' such that $\theta_{j'} = f$ and such that

$$g'(x, S_{j'}(x)) \le S_{i'}(x) \text{ a.e.}$$

Then since t is onto, if $\phi_i = f$ there is a j such that $\phi_j = f$ and such that

$$g'(x, S_{t(j)}(x)) \le S_{t(i)}(x) \text{ a.e.}$$

Therefore,

$$r(x, g(x, \Phi_j(x))) \le r(x, g(x, r(x, S_{t(j)}(x)))) \le S_{t(i)}(x) \le r(x, \Phi_i(x)) \text{ a.e.}$$

This, together with the fact that r is nondecreasing in its second argument, yields

$$g(x, \Phi_j(x)) \le \Phi_i(x) \text{ a.e.,}$$

which completes the proof of the Blum Speedup Theorem. $\square$

The Blum Speedup Theorem puts an end to any hope of having ''best'' programs for *all* computable functions. It shows that for any total recursive g there is a (small) total recursive function f for which there is no best program to within a factor of g. Thus, we cannot in general assign to a computable function f another computable function C_f which we could call the "complexity of f" in the sense that it is a best "running time" for programs which compute f.]

Finally, as is the case with the Gap Theorem, there is a provocative interpretation of the Blum Speedup Theorem: for computing some total recursive functions a fast computer is no better than a slow computer! This time, suppose we have two complexity measures which are related by the recursive function r, and consider a function f with r-speedup in the "slower" measure; then for every program which computes f in the "fast" measure there is *some* program which computes f just as "fast" in the slower measure.

Additional Exercises

*5.3.5

(a) Show that for our RAM programming system and space measures and for the TM*space* measure there is a primitive recursive s-1-1 function s such that for all i, x, and y, $S_{s(i,x)}(y) = S_i(x, y)$. *Hint*: do not try to track back through the definition of s-m-n functions from functions for composition (for arbitrary acceptable programming systems) in Chapter 3 and then through the definition of the function for composition in the RAM system in Chapter 2; instead, define s directly using programming and coding in the RAM system and then in the Turing machine system.

(b) Also show that all of these measures satisfy condition (b).

5.3.9 Show that the function t in the Gap Theorem can be made arbitrarily large; that is, let g be any total recursive function such that $y < g(x, y)$ for all x and y, let b be any total recursive (bounding) function, and show that there is a total recursive function t such that $b(x) \leq t(x)$ for all x and such that if $t(x) < \Phi_i(x) < g(x, t(x))$ then $x \leq i$.

5.3.10 Show that there are arbitrarily complex *characteristic functions*; that is, show that for every total recursive function t there is a total recursive characteristic function f such that if $\phi_i = f$, then $\Phi_i(x) \geq t(x)$ a.e.

5.3.11 Show that there are arbitrarily complex *sparse* characteristic functions; that is, show that for every total recursive function t there is a total recursive characteristic function f such that if $\phi_i = f$ then $\Phi_i(x) \geq t(x)$ a.e. and such that for all x,

$$(\Sigma_{y \leq x} f(y))/(x + 1) \leq .04.$$

(The values of such a function f are very hard to compute "for sure," but very easy to "guess" with high probability of success!)

5.3.12 Prove the Blum Speedup Theorem directly, without using the recursive relatedness of complexity measures.

5.3.13 Show that the function f in the Blum Speedup Theorem can be taken to be a characteristic function; that is, replace the condition "$f(x) \leq x$" by "$f(x) \leq 1$" in the statement of the theorem, and prove this modified statement.

5.4 GENERAL SIZE MEASURES FOR PROGRAMS

In this section we explore very briefly what happens when, in addition to measuring the amount of a computational resource a program uses in its computations, we add a measure of program size. We give results which indicate the "futility of optimization," and which indicate what sacrifices one must be prepared to make in return for the benefits, if any, of programming in a "restricted" programming system rather than in a general purpose programming system which can compute all of the partial recursive functions.

Throughout this section, ϕ_0, ϕ_1, ... and Φ_0, Φ_1, ... stand for any acceptable programming system and any general complexity measure on that system. We define a general *size measure* on programs to be any total recursive function s such that for all n, $\{\phi_i : s(i) = n\}$ is finite; that is, we require that given a program we be able to effectively compute its size, and that the number of *inequivalent* programs of any given size be finite. We usually denote $s(i)$, the size of program i, by the more suggestive notation $|i|$.

Counting the total number of characters in a RAM program gives a size measure, and any reasonable way of figuring the size of programs in a "natural" programming system gives a size measure, since there will be only finitely many *distinct* programs of any given size. Our definition also allows some slightly unreasonable ways of counting as size measures; for example, the number of instructions in a RAM program or in a Turing machine (over some fixed alphabet) give size measures. There are actually infinitely many *distinct* programs of any given size, but only finitely many *inequivalent* ones since changing the "names" of the registers or of the states does not "really" change the program. Note however, that the number of instructions in an ALGOL program is not a size measure since a single instruction can compute any constant function.

As is the case with complexity measures, our definition of size measures is sufficiently weak to allow many "pathological" size measures along with all of the natural size measures. For example, if p is a padding function for our programming system (see Proposition 3.4.5) there is nothing to prevent a size measure in which $|p(i, x)| = |i|$ for all i and x (so no matter how much we "pad" program i, it never gets any longer!); in fact, if we take any size measure we can simply redefine it on the range of p so that this is the case (provided the range of p is recursive, which is easily arranged).

5.4.1 EXERCISE Show how to replace any padding function p by a padding function p' satisfying $p'(i, x) > max\{i, x\}$. Note that the range of p' is recursive, and then show how to construct a size measure such that $|p'(i, x)| = |i|$ for all i and x.

In spite of the weakness of our definition of size measures, there are still some very interesting and important results we can prove for *all* size measures. We begin by using a helpful technique which shows that in spite of the example we have just discussed, with any size measure we can still pad programs.

5.4.2 PROPOSITION (Length Padding Lemma) *For any size measure there is a total recursive function p such that for all i, $\phi_i = \phi_{p(i)}$ and $|i| < |p(i)|$.*

Proof Let the program i be given, and we shall show how to compute $p(i)$. For each natural number y define a total recursive function f (using our standard s-m-n construction) such that for *all* n and x

$$\phi_{f(n)}(x) = \begin{cases} \phi_i(x) & \text{if } |n| > |i| \\ y & \text{if } |n| \le |i|. \end{cases}$$

Note that f depends on both i and y, and that a program for f can be found effectively from i and y, but for the sake of notational simplicity we are not explicitly displaying this. Now let n be a fixed program such that $\phi_n = \phi_{f(n)}$. Since a program for f can be found effectively from i and y, and since n can be found effectively from such a program by the Extended Recursion Theorem (3.4.2), given i and y we can effectively compute n (which depends on i and y). If $|n| > |i|$ then we can define $p(i) = n$ since $\phi_n = \phi_{f(n)} = \phi_i$, but we may be unlucky and find that $|n| \le |i|$. However, in this case we have not totally wasted our time since we now have the information that n is a program for the constant function y whose length is no more than $|i|$.

Now suppose we perform the procedure above with $y = 0$, then with $y = 1$, etc., searching for an n such that $|n| > |i|$ and hence such that $\phi_n = \phi_i$. This search cannot go on forever; each time a new value of y

produces an n no longer than $|i|$ we have a program for a *different* (constant) function, and there are only finitely many functions with programs for computing them whose length is no greater than $|i|$. Therefore, our procedure applied with successively larger values of y *must* eventually produce an appropriate value for $p(i)$. (Make sure you understand this proof because we use the same construction in a slightly more complicated context in the proof of the next theorem.) $\square$

When computer scientists speak of "optimizers" they generally do not intend the word to be taken literally, but instead they are generally referring to algorithms which take programs (sometimes of a very restricted type) as inputs and output equivalent programs which, in "many cases," are hopefully somewhat "better." Let us call a total recursive function f a *preprocessor* if $\phi_{f(i)} = \phi_i$ for all i; that is, if f preserves the equivalence of programs. Our next theorem shows that *every* preprocessor must be very far indeed from being an "optimizer" in the literal sense of the word.

5.4.3 THEOREM   *Let f be any preprocessor, and let g and h be any (intuitively "large") total recursive functions. Then for every program i we can effectively find a program j such that*

1. $\phi_j = \phi_{f(j)} = \phi_i$
2. $|f(j)| \geq g(|i|)$
3. $\Phi_{f(j)}(x) \geq h(x, \Phi_i(x))$ for all x.

(Intuitively, this says that for every program i we can find an equivalent program j such that its "improved" version $f(j)$ is far from "optimal" because it has both much greater length than i and much greater "running time" than i; thus f redefined to have $f(j) = i$ for each such j would be a much "better" preprocessor, but still far from an "optimizer." Such a result is not surprising once one realizes that the range of any preprocessor will still be a programming system and therefore should contain some very bad programs, as shown in Exercise 5.4.5.)

Proof   Let the program i be given. Then for any program n we can use essentially the same construction as in the proof of the previous proposition to effectively find a program j such that $\phi_j = \phi_n$ and $|f(j)| \geq g(|i|)$, where the program j depends (effectively) on i and n, but for notational simplicity we do not explicitly display this. The construction uses a total recursive function m (which depends on i, n, and y) such that

$$\phi_{m(z)}(x) = \begin{cases} \phi_n(x) & \text{if} \quad |f(z)| \geq g(|i|) \\ y & \text{if} \quad |f(z)| < g(|i|) \end{cases}$$

for all z and x. Then taking fixed points for the functions m for successively larger values of y will eventually produce the required j.

5.4.4 EXERCISE Verify the previous sentence.

We now continue with the proof of the theorem. For each n we have a j which depends effectively on i and n such that $\phi_j = \phi_n$ and $|f(j)| \geq g(|i|)$. Then we define a total recursive function t such that for all n and x,

$$\phi_{t(n)}(x) = \begin{cases} \phi_i(x) & \text{if} \quad \phi_i(x) \, convergent \, and \, \Phi_{f(j)}(x) \geq h(x, \Phi_i(x)) \\ divergent & \text{otherwise.} \end{cases}$$

We now take n as a fixed program such that $\phi_{t(n)} = \phi_n$. Note that since a program for t can be found effectively from i, n can be found effectively from i, and thus the j which depends on i and n can then be found effectively from i; from now on, we use j to stand for that particular value of j which depends on i and the fixed point n.

We know that $\phi_{f(j)} = \phi_j = \phi_n = \phi_{t(n)}$. For any x, if $\phi_i(x)$ is divergent then $\phi_{t(n)}(x) = \phi_j(x)$ is also divergent, and so

$$\Phi_{f(j)}(x) = \infty \geq h(x, \Phi_i(x)) = \infty$$

by our convention (and of course, $\phi_i(x) = \phi_j(x)$). If $\phi_i(x)$ is convergent then $\phi_{t(n)}(x)$ *cannot* be divergent because if it were we would have $\Phi_{f(j)}(x) = \infty \geq h(x, \Phi_i(x))$, which would yield $\phi_{t(n)}(x) = \phi_i(x)$ by the definition of t. Therefore, $\phi_i = \phi_j = \phi_{f(j)}$ and $\Phi_{f(j)}(x) \geq h(x, \Phi_i(x))$ for all x, which completes the proof of the theorem. $\square$

If we take f to be the identity function in the previous theorem, the theorem shows that for any acceptable programming system, complexity measure, and size measure, we can effectively find arbitrarily "bad" programs equivalent to any given program. (Observing a beginning programming class provides experimental verification of this fact.)

5.4.5 EXERCISE Give a precise formulation of the remark we have just made.

Our next goal is to explore what happens if we no longer require f to be a preprocessor, that is, to produce programs equivalent to its inputs, but simply let f be any effective listing of programs. We would like to show that any effective listing of programs must contain some very bad programs. But this is obviously not true: f could list one optimal program, or infinitely many different but trivially equivalent optimal programs. Thus we shall have to require f to list infinitely many inequivalent programs, or at least infinitely many programs of different lengths (which is a weaker requirement), before we can hope to show such a result. But even if f lists infinitely many inequivalent programs, we must still be circumspect in how strong a meaning we try to give to the term "bad program"; in most "natural" measures we can effectively list very "efficient" programs for computing the constant functions. Nevertheless, such a listing must contain some programs which are much longer than they need to be; this is made precise in Theorem 5.4.7. However, first we need a further, "complexity-theoretic" extension of the Recursion Theorem.

Recall, in case you could possibly have forgotten, that the Recursion Theorem (3.4.1) asserts that for every total recursive function f there is a fixed point program n such that $\phi_n = \phi_{f(n)}$. At first glance it may appear that the program $f(n)$ uses the program n as a subroutine in such a way that we might expect that $\Phi_n(x) \leq \Phi_{f(n)}(x)$. However, this conclusion is not warranted; although $f(n)$ *probably* uses n as a subroutine, it *may very well* use it on arguments other than x in computing $\phi_{f(n)}(x)$. If you work Exercises 5.4.11 and 5.4.12 you will see specific examples of why $\Phi_n(x)$ cannot be bounded by $\Phi_{f(n)}(x)$.

At this point it may seem that there is no coherent relationship between Φ_n and $\Phi_{f(n)}$. However, a careful examination of the proof of the Recursion Theorem shows that the program n actually simulates the program $f(n)$, and so we can hope to bound Φ_n in terms of $\Phi_{f(n)}$ by estimating the overhead for this simulation. Although we do not use this idea in the proof of the Complexity-Theoretic Recursion Theorem below, we heartily encourage the reader to ponder the proof of the Recursion Theorem and get a firm intuitive grasp of what we have just said since it would be helpful for a clear understanding of Chapter 6.

5.4.6 COMPLEXITY-THEORETIC RECURSION THEOREM   *There are total recursive functions n and h such that if ϕ_i is total then*

$$\phi_{\phi_i(n(i))} = \phi_{n(i)} \quad \text{and} \quad \Phi_{n(i)}(x) \leq h(x, \Phi_{\phi_i(n(i))}(x)) \text{ a.e.;}$$

moreover, $h(x, y) \leq h(x, y + 1)$ for all x and y.

Proof   The function n is supplied by the Extended Recursion Theorem (3.4.2), and the remainder of this proof works for any such fixed point function. Once again, we are going to use a "max-ing" construction as we did in the proofs of the Recursive Relatedness and Compression Theorems. We define a total recursive function k such that for all i, x, and y,

$$k(i, x, y) = \begin{cases} \Phi_{n(i)}(x) & \text{if } \Phi_i(n(i)) \leq x \text{ and } \Phi_{\phi_i(n(i))}(x) = y \\ 0 & \text{otherwise.} \end{cases}$$

Observe that if ϕ_i is total (in fact, just if $\phi_i(n(i))$ is convergent), then

$$k(i, x, \Phi_{\phi_i(n(i))}(x)) = \Phi_{n(i)}(x)$$

for $x \geq \Phi_i(n(i))$ if $\phi_{\phi_i(n(i))}(x)$ is convergent since $\phi_{\phi_i(n(i))} = \phi_{n(i)}$. On the other hand, if $\phi_{\phi_i(n(i))}(x)$ is divergent then

$$\Phi_{n(i)}(x) = \Phi_{\phi_i(n(i))}(x) = \infty.$$

Now we define $h(x, y) = max_{i \leq x} max_{z \leq y} k(i, x, z)$ for all x and y. Then certainly $h(x, y) \leq h(x, y + 1)$ for all x and y, and if ϕ_i is total then

$$\Phi_{n(i)}(x) = k(i, x, \Phi_{\phi_i(n(i))}(x)) \leq h(x, \Phi_{\phi_i(n(i))}(x))$$

for all $x \geq max\{i, \Phi_i(n(i))\}$. $\square$

Given the way h was constructed in the previous proof, there is nothing very enlightening we can say about its size, except that for the particular acceptable programming system and complexity measure it does give a *fixed* recursive bound on the complexity of a fixed point program n in terms of the complexity of the program $f(n)$ for all total recursive functions f. As we intimated in the discussion just prior to the statement of the theorem, by examining the computation of the fixed point program n in any "natural" programming system under any "natural" complexity measure, it is possible to see just how large h needs to be. The possibly surprising, and definitely important, fact is that in most "natural" programming systems under most "natural" complexity measures, h can be taken to be a *polynomial* of small degree; and for many measures, with very careful programming it can be shown that h can be simply a *multiplicative constant* which depends on the size of the program n. You should read the proof of Theorem 6.1.8 and do Exercise 6.1.12 to see how this is done.

We now use the Complexity-Theoretic Recursion Theorem to show that every effective listing containing arbitrarily long programs must contain some programs which are excessively long and which could be replaced by much shorter equivalent programs whose complexity is not much greater; the bound on this increase in complexity is the function h from the previous theorem, and we repeat that for "natural" measures this bound is very small indeed, often resulting in a "mere" linear increase in complexity.

5.4.7 THEOREM *For any total recursive function f such that the range of $|f|$ is infinite (i.e., f lists programs of infinitely many different sizes), and for any total recursive function g, we can effectively find a* *program i and an integer j such that*

1. $\phi_i = \phi_{f(j)}$
2. $g(|i|) \leq |f(j)|$
3. $\Phi_i(x) \leq h(x, \Phi_{f(j)}(x))$ a.e.

where h is the total recursive function from the Complexity-Theoretic Recursion Theorem.

Proof   Since the range of $|f|$ is infinite, we can define a total recursive function k by $k(y) = f(min\ z[g(|y|) \leq |f(z)|])$; that is, $k(y)$ is the first program f lists which has length at least $g(|y|)$. From the Recursion Theorem take i such that $\phi_i = \phi_{k(i)}$ and let $j = min\ z[g(|i|) \leq |f(z)|]$ so that $f(j) = k(i)$. Then $\phi_i = \phi_{k(i)} = \phi_{f(j)}$ and $g(|i|) \leq |f(j)|$, and by the Complexity-Theoretic Recursion Theorem,

$$\Phi_i(x) \leq h(x, \Phi_{k(i)}(x)) = h(x, \Phi_{f(j)}(x))\ \text{a.e.}$$

which completes the proof of the theorem.   □

   This theorem can be used to show that many "optimization" problems do not have effective solutions. For example:

5.4.8 EXERCISE

   (a) Define the total function f for all i by

$$f(i) = min\ j[\phi_j(0) = i\ and\ \forall k(\phi_k(0) = i\ implies\ |k| \geq |j|)];$$

   $f(i)$ is just the shortest program which "prints" i. Show that f cannot be a recursive function. (This is just the generalization to any acceptable programming system and size measure of the result from the Introduction which was restated for RAM programs as Proposition 5.1.11.)
   (b) Show that the set M of "minimal" programs is not r.e.; that is, show that

$$M = \{i : \forall j(\phi_j = \phi_i\ implies\ |j| \geq |i|)\}$$

   is not r.e. You may want to review the basic properties of recursively enumerable sets.

   Theorem 5.4.7 can also be used to gain information about restricted programming languages. It might be, and sometimes is, argued that a general purpose programming language suffers from the inability to decide much of anything about programs in the language because it is *too* general. If we could somehow restrict ourselves to using a much more manageable set of programs, bothersome phenomena such as the unsolvability of the halting problem and speedup might not occur;

indeed this is certainly true, but few programmers are willing to use *programs* only of the form **print** n. Still, there are very many possible intermediate proposals which merit some attention. The rest of this section is devoted to two results about what happens when one somehow restricts the set of programs so that they all compute total recursive functions; of course, Proposition 3.3.6 shows that any (reasonable) such restricted programming language will not compute all of the total recursive functions, but it still may compute all of those we ever actually attempt to compute in practice. The results in Section 1.10 indicate that any function anyone might actually want to attempt to compute is certainly primitive recursive, and in Chapter 1 we examined a restricted programming language which computes (exactly) the primitive recursive functions. Our final two results of this section apply to such restricted programming languages.

We say that a recursive set RPS is a *restricted programming system* if $\{\phi_i : i \in \text{RPS}\}$ is an infinite set of total recursive functions. That is, a restricted programming system is simply any recursive set of programs for an infinite collection of total recursive functions.

For our results on restricted programming systems we are going to require that the size measure on programs satisfy a stronger condition than just the (very weak) definition given earlier. We say that a size measure is *canonical* if there is a total recursive function b such that for all i and j, if $|i| \leq j$ then $i \leq b(j)$; that is, $b(j)$ gives us a *computable* bound on the *finite* set of programs which can have length no greater than j. This strengthens the definition of size measures in two ways: first, there can only be finitely many *programs* of any given size; and second, for any given size j we can *effectively* find all programs of that size simply by listing $\{i : i \leq b(j) \text{ and } |i| = j\}$. Any reasonable way of measuring the size of programs in a "natural" programming system will certainly yield a canonical size measure.

5.4.9 PROPOSITION *Suppose we have a canonical size measure and that* RPS *is a restricted programming system, then the set of "minimal" programs in* RPS *is recursively enumerable. Specifically, there is a total recursive function f whose range is contained in* RPS *such that*

$$\{\phi_{f(i)} : i \in N\} = \{\phi_i : i \in \text{RPS}\}$$

and

$$\phi_k = \phi_{f(i)} \text{ for } k \in \text{RPS} \quad \text{implies} \quad |k| \geq |f(i)|.$$

Proof Since the size measure is canonical and RPS is recursive, for any given program in RPS we can effectively find the finite set of programs in RPS which are shorter; this, together with the fact that all of the programs in RPS compute *total* functions, shows that ψ defined as follows is a partial recursive function:

$$\psi(x) = \begin{cases} min \ z[\forall y(y \in RPS \ and \ |y| < |x| \ implies \\ \qquad \exists w \leq z(\phi_x(w) \neq \phi_y(w)))] & \text{if } x \in RPS \\ divergent & \text{if } x \notin RPS. \end{cases}$$

Then the domain of ψ is the set of minimal length programs in RPS and Proposition 3.3.2 provides the total recursive function f which enumerates this set. $\square$

The previous proposition, taken in contrast with Exercise 5.4.8b, shows that restricted programming systems can offer some improvements over (general) acceptable programming systems. On the other hand, our next theorem shows that restricted programming systems always compute some functions whose only programs in the restricted language are excessively long.

5.4.10 THEOREM   *Suppose that we have a canonical size measure, that RPS is a restricted programming system, and h is the function from the Complexity-Theoretic Recursion Theorem. Then for any total recursive function g we can effectively find programs i and j such that*

1. $i \notin RPS$, $j \in RPS$, and $\phi_i = \phi_j$
2. $|j| \geq g(|i|)$
3. $\forall k(k \in RPS \ and \ \phi_k = \phi_i \ implies \ |k| \geq |j|)$
4. $\Phi_i(x) \leq h(x, \Phi_j(x))$ a.e.

Proof   This theorem follows directly from Theorem 5.4.7 and Proposition 5.4.9. $\square$

Additional Exercises

5.4.11 Consider our RAM programming system and pseudospace measure, and let S_j be any (fixed) total pseudospace function. Show that there is a primitive recursive function f such that for all i and x,

$$\phi_{f(i)}(x) = \begin{cases} 0 & \text{if } S_j(x) < S_i(x) \\ divergent & \text{otherwise,} \end{cases}$$

and such that if $\phi_{f(i)}(x)$ is convergent then $S_{f(i)}(x) = S_j(x)$. Let n be a fixed point program for f; that is, $\phi_n = \phi_{f(n)}$. Show that ϕ_n is total, and that $S_{f(n)}(x) < S_n(x)$ for all x.

5.4.12

(a) Let f be a total recursive function such that for all i, $\phi_{f(i)}(0) = 0$ and $\phi_{f(i)}(x) = \phi_i(x/2)+1$ for all $x \geq 1$; let n be a fixed point program such that $\phi_n = \phi_{f(n)}$. Give, and prove correct, a simple description of the (total) function ϕ_n.

Consider Turing machines with alphabet $\{0, 1, B\}$ (so numbers are represented in binary) which have three separate tapes (and heads): a work tape, a read-only input tape, and a one-way, write-only output tape which moves from left to right (so that the least significant bits of outputs are written first), and let the "space" measure be the number of squares used on the work tape.

(b) Explain why, in the space measure just given, for every y there must be an x such that $y < \Phi_n(x)$; that is, Φ_n is unbounded.

(c) Assuming f is defined in the "natural" way, explain why $\Phi_{f(n)}(x) < \Phi_n(x)$ i.o. in the space measure just given.

5.4.13 Prove the following extension of the ordinary padding lemma: given any nonempty recursive set of equivalent programs, we can effectively find another equivalent program not in the set. Specifically, show that for any acceptable programming system $\phi_0, \phi_1, \ldots$ there is a total recursive function p such that for all j, if ϕ_j is a total characteristic function such that for some i, $\phi_j(i) = 1$ and $\phi_j(x) = 1$ implies $\phi_x = \phi_i$, then $\phi_{p(j)} = \phi_i$ and $\phi_j(p(j)) = 0$.

5.4.14 Select several of the space measures of Section 1.9 and prove that for these measures the function h of Theorem 5.4.6 can be replaced by $\Phi_{n(i)}(x) \leq k_i[\Phi_{\phi_i(n(i))}(x)]$ where k_i is a constant which may depend on i. (Go back and look at the original proof of the Recursion Theorem 3.4.1. This exercise holds for some time measures but may require tricky programming.)

Chapter 6

Exponentially and Superexponentially Difficult Problems

In the previous chapter we developed a very general theory of computational complexity which, partially because it is insensitive to the properties of particular computing devices, programming languages, and conventions for representing inputs and outputs, gives general results which hold for virtually any conceivable general programming system and complexity measure. Unfortunately however, that theory does not yield any useful information about the computational complexity of specific computational problems (partial recursive functions). In this and the following chapter we develop less general theories which do yield information on the complexity of specific computational problems in large classes of "natural" complexity measures.

In the first section of this chapter we develop a general method for proving specific lower bounds on the complexity of particular computational problems in specific complexity measures. This general method is based on the intuition that, essentially, the only general way to tell whether or not an arbitrary program on an arbitrary input uses more than a given amount of some computational resource is simply to run the program and see, and that doing this uses at least the given amount of the computational resource.

In Section 6.2 we use the general method of Section 6.1 to show that an algorithmically solvable problem concerning an extended set of regular expressions requires space exponential in the length of the inputs for infinitely many inputs for *any* Turing machine which solves the problem. Since, to within a small polynomial factor, Turing machine space measure serves as a lower bound for all "natural" complexity measures which people have considered, this shows that the problem has at least exponential complexity in all such measures. The problem is therefore intractable in the sense that there is no practical algorithm for solving it.

In Section 6.3 we show that an algorithmically solvable problem for another set of extended regular expressions has superexponential (that is, nonelementary) complexity. In the optional Section 6.4 we show that the theory of addition on the natural numbers has more than exponential complexity in the MIN*space* measure and conclude that this theory has at least exponential complexity in the Turing machine time measure. In

168

the final optional section we give a decision procedure for this theory. Thus the theory of addition on the natural numbers, although decidable, has no practical decision procedure.

6.1 LINEARLY BOUNDED COMPLEXITY MEASURES AND LIMITED HALTING PROBLEMS

In earlier chapters we saw that many interesting and natural questions which arise in computer science and mathematics have no algorithmic solutions. In this chapter we present tools which make it possible to prove that some problems, although algorithmically solvable, are *intractable* in the sense that any algorithm for solving them must use excessive amounts of computational resources. In particular, in this section we develop a general method which can be used to show that specific problems are at least exponentially difficult.

When dealing with complexity measures on acceptable programming systems in this chapter we shall (generally) think of the arguments to functions as words over some alphabet A_k with $k \geq 1$ or else as n-tuples of words over such alphabets. In other words, we shall use *natural* complexity measures, such as those in Section 1.9. While formally $\phi_i(x,y)$ will still stand for the result of running program i on the integer input $\langle x,y \rangle$, the complexity $\Phi_i(x,y)$ will be determined by running program i on the pair of input words x and y in A_k^* with $k \geq 1$. For example, if we are considering the Turing machine space measure TM*space*, then for each Turing machine program i there will be a number n of arguments associated with it, and the complexity $\Phi_i(x)$ of $\phi_i(x)$ will be the space used by that Turing machine on the inputs $D_k(\Pi(1, n, x)), \ldots, D_k(\Pi(n, n, x))$. Since these *words* are the actual inputs given to the Turing machine, TM*space* is the appropriate space measure for the Turing machine programming system.

Since in this chapter we wish to measure the use of computational resources as a function of the length of the inputs, we use $|x|$ to denote the length of the word $D_k(x)$ encoding x. If we are thinking of x as coding an n-tuple $w_1, \ldots, w_n$ of words, then we use the notation $\bar{x}$ to indicate this and $|\bar{x}| = \Sigma_{1 \leq i \leq n} |w_i|$. However sometimes, as in Section 6.4, it will be more convenient simply to take $|x|$ to be the integer x. For the results of this section, and in particular for Theorem 6.1.9, it does not matter how the length function is defined.

Now of course we know from Proposition 5.2.2 that given any algorithmically solvable problem an unnatural complexity measure can be created in which that specific problem has zero complexity; thus we cannot expect to prove that any specific problem is intrinsically difficult in *all* complexity measures. We also know from the Compression Theorem (5.3.3) that in any complexity measure there are (fairly small) arbitrarily complex total recursive functions, but these functions are "artificial" in the sense that they were constructed specifically to be difficult to compute. They did not arise from a "natural" computational problem of independent interest. In this and the following chapter we are interested in showing that some specific computational problems of independent interest are intrinsically difficult, and to do so we must put some restrictions on the complexity measures we use.

From Section 1.9 we know that the translations among the programming systems given in Chapter 1 have the property that the complexity of a translated program can be bounded by the composition of a reasonable function (usually a small degree polynomial) with the complexity of the original program. Moreover, there are translations among all natural programming systems, including all programming languages actually in use, with similarly reasonable bounds. Thus, if it is proven that a computational problem has at least exponential complexity in, for example, the Turing machine *space* measure, then it has exponential complexity in all reasonable complexity measures, including those considered in Section 1.9. This is the approach to exponential complexity we use in this chapter.

In previous chapters we saw that the halting problem is a basic unsolvable problem in acceptable programming systems, and this fact was used to show that many other problems are unsolvable; from this it is natural to arrive at the intuition that a limited halting problem should be an intrinsically difficult computational problem. Thus if t is a total recursive function and we wish to show that any algorithm for determining membership in some (recursive) set S has complexity at least $t(|x|)$ on infinitely many inputs x, we might try to find words $R(i, x)$ such that $R(i, x) \in S$ if and only if $\Phi_i(x) > t(|x|)$. If the words $R(i, x)$ are reasonably short and the complexity of producing them as a function of i and x is reasonably small, then the complexity of testing membership in S should not be much less than t, for if it were there would be a general way to test whether $\Phi_i(x) > t(|x|)$ with complexity much less than t simply by testing whether $R(i, x)$ is in S. *Intuitively*, there should be no general way to test whether $\Phi_i(x) > t(|x|)$ which has complexity less than t since the only general way to make such a test should be to compute $\phi_i(x)$ and see, but it is by no means obvious that this intuition is correct.

By definition, every acceptable programming system $\phi_0, \phi_1, \ldots$ has a total recursive function c for composition such that for all i and j, $\phi_{c(i,j)} = \phi_i \circ \phi_j$. In Section 3.1 we showed that every acceptable programming system also has a total recursive s-1-n function s such that

$\phi_{s(i,x)}(\vec{y}) = \phi_i(x, \vec{y})$ for all i, x, and $\vec{y}$. By a straightforward use of the standard s-1-n construction, every acceptable programming system also has a total recursive function c' for "extended" composition such that $\phi_{c'(i,j)}(\vec{x}, \vec{y}) = \phi_i(\phi_j(\vec{x}), \vec{y})$ for all $i, j, \vec{x}$, and $\vec{y}$.

6.1.1 EXERCISE   Verify the assertion we have just made.

Because it is always clear from the context which of the functions c and c' is meant, for the remainder of this section we shall adopt the notational convenience of using c to stand for both the functions c and c' above. Thus, c will stand for a total recursive function such that $\phi_{c(i,j)}(\vec{x}, \vec{y}) = \phi_i(\phi_j(\vec{x}), \vec{y})$ for all $i, j, \vec{x}$, and $\vec{y}$, and when c stands for the original composition function $\vec{y}$ will be a "0-tuple" (that is, $\vec{y}$ is not really there), while when c stands for the extended composition function $\vec{y}$ will be an n-tuple with $n \geq 1$.

The s-1-n and composition functions exist for all acceptable programming systems. Moreover, as was discussed following the proof of the s-m-n Theorem (3.1.2), in *natural* programming systems, s-1-n and extended composition functions can be constructed *directly* by describing the very simple manipulations on programs needed to obtain the desired results. When this is done, the complexity of the manipulated programs in any natural complexity measure will be bounded by the composition of very simple functions with the complexities of the original programs. Thus, we make the following definition.

6.1.2 DEFINITION   A complexity measure Φ_0, Φ_1, ... on an acceptable programming system ϕ_0, ϕ_1, ... is *linearly bounded* if $\Phi_i(\vec{x}) \geq |\vec{x}|$ for all i and $\vec{x}$, and if there are a total recursive s-1-n function s and total recursive composition functions c for the system and there is a positive integer constant k such that for all i, j, and z

1. $\Phi_{c(i,j)}(\vec{x}, \vec{y}) \leq k[\Phi_i(\phi_j(\vec{x}), \vec{y}) + \Phi_j(\vec{x})]$ a.e. $\vec{x}, \vec{y}$
2. $\Phi_{s(i,z)}(\vec{x}) \leq k[\Phi_i(z, \vec{x})]$ a.e. $\vec{x}$.

Because of the very simple manipulations of programs performed by (properly chosen) functions c and s, most natural complexity measures on natural programming systems are linearly bounded, including most of the complexity measures considered in Section 1.9. Note that the difficulty of computing c and s themselves does not matter.

To verify that a complexity measure is linearly bounded you should assume that the programs have been numbered in some reasonable way similar to our coding of RAM programs in Section 2.4 and you should not worry about explicitly defining appropriate functions c and s, but simply concern yourself with the manipulations of programs these functions would describe and what effect on complexities these manipulations would have. Keeping this in mind, proving that the complexity of composition is appropriately bounded is usually very straightforward, and the same is true of the complexity of the programs produced by the s-1-n function, particularly once one remembers that the program $s(i, z)$ should simply copy the value z into the input of program i and then transfer control to program i.

For s-1-n functions, the transfer of control is easily represented by a composition and so Condition 2 often "very nearly" follows from Condition 1 in two different ways. Once Condition 1 has been verified, to verify Condition 2 it is sufficient to prove that there is some total recursive function $pair$ such that $\phi_{pair(j)}(x) = (j, x)$ and for each j

$$\Phi_{pair(j)}(x) \le k \cdot |(j, x)| \quad \text{a.e.}\ x$$

where k is the constant from Condition 1. Alternatively, if there is a slightly different composition function c' such that for all x

$$\phi_{c'(i,j)}(x) = \phi_i(\phi_j(x), x)$$

and such that

$$\Phi_{c'(i,j)}(x) \le k \cdot [\Phi_i(\phi_j(x), x) + \Phi_j(x)] \quad \text{a.e.}\ x,$$

then it is adequate to have any total recursive function $const$ such that for each j $\phi_{const(j)}(x) = j$ for all x and such that

$$\Phi_{const(j)}(x) \le k \cdot |(j, x)| \quad \text{a.e.}\ x.$$

6.1.3 EXERCISE

(a) Verify that the existence either of an appropriate function $const$ or of an appropriate function $pair$ as described above (along with the appropriate composition functions c) is adequate to establish the appropriate s-1-n function of Condition 2 of Definition 6.1.2.

(b) Verify that all of the measures of Section 1.9 are linearly bounded, with the possible exception of TM$time$.

The notion of linearly bounded complexity measures gives a class of nicely behaved measures for turning our earlier intuition about limited halting problems into a proof. However, we still need a formal definition of when a (recursive) set S is rich enough to express a limited halting problem. In the definition below, r is a program for producing the words $R(i, x)$ such that $R(i, x) \in S$ if and only if $\Phi_i(x) > t(|x|)$; that is, $\phi_r(i, x) = R(i, x)$.

6.1.4 DEFINITION Let $\phi_0, \phi_1, \ldots$ be an acceptable programming system and let $\Phi_0, \Phi_1, \ldots$ be a complexity measure on it. For any total recursive function t, a (recursive) set S is *rich enough to express the*

t-*limited halting problem* for this programming system and measure if there is a program r such that for all i, $\phi_r(i, x)$ is one-to-one as a function of x, and such that

1. $\phi_r(i, x) \in S$ iff $\Phi_i(x) > t(|x|)$ for all i and x
2. $|\phi_r(i, x)| \leq k_i \cdot |x|$ a.e. x, where k_i is an integer constant depending on i
3. $\Phi_r(i, x) \leq t(|\phi_r(i, x)|/k)$ a.e. x, for all positive integers k.

Moreover, S is rich enough to *strongly* express the t-limited halting problem if in Condition 2 the constants k_i can be replaced by a single constant k independent of i.

In the definition above, Conditions 2 and 3 make precise what is meant by requiring $R(i, x)$ to be reasonably short and what is meant by requiring the complexity of producing $R(i, x)$ to be reasonably small. Notice that if t is a large function, Condition 3 allows the complexity of producing $R(i, x)$ to be almost as large as t.

As a simple illustration of the ideas above, for any total recursive function t define the set $S_t = \{\langle i, x \rangle : \Phi_i(x) > t(|x|)\}$. If the pairing function "$\langle , \rangle$" is reasonable and if the identity function is sufficiently easy to compute in the complexity measure Φ, then S_t is obviously rich enough to express the t-limited halting problem for that measure. S_t is a recursive set, and intuitively, any program d for the characteristic function of S_t should satisfy $\Phi_d(y) \geq t(|y|)$ i.o. You might try to prove this now. In fact, one expects $\Phi_d(y)$ to be roughly equal to $t(|y|)$.

Before we can finally turn our intuitions about t-limited halting problems into a proof, we need a few more simple definitions and a simple fact about linearly bounded complexity measures. Recall that in Section 3.3 we discussed recursively enumerable sets as the class of partially solvable problems and showed that the recursively enumerable sets are the domains of the partial recursive functions. If S is a (recursive) set then a partial recursive function ψ is a *partial decision procedure* for S if S is the domain of ψ. In any natural programming system, a program which decides membership in a recursive set S (that is, a program for the characteristic function of S) can be very easily converted into a program for a partial decision procedure for S; moreover, with any natural complexity measure the complexity of the converted program on members of S will be very little more than the complexity of the original program on members of S. We can use the following proposition to make this precise.

6.1.5 PROPOSITION *Let* Φ_0, Φ_1, ... *be a linearly bounded complexity measure on the acceptable programming system* ϕ_0, ϕ_1, ..., *and let p be a program such that* $\phi_p(1) = 1$ *and* $\phi_p(x)$ *is divergent if* $x \neq$

1. *Then there is a positive integer k' such that for all j, $\Phi_{c(p,j)}(x) \leq k'[\Phi_j(x)]$ for all sufficiently large x such that $\phi_j(x) = 1$, where c is a total recursive function for composition as in the definition of linearly bounded measures.*

6.1.6 EXERCISE   Prove the previous proposition.

Notice that in the previous proposition, if ϕ_j is the characteristic function of a set S then $\phi_{c(p,j)}$ is a partial decision procedure for S; thus if S has only complex *partial* decision procedures, it can have only complex decision procedures.

Finally, we say that a total function t is *at least linear* if $t(x) \leq t(x+1)$ for all x and if for all positive integers k, $k \cdot t(x) \leq t(k \cdot x)$ a.e. For the proof of our main theorem, Theorem 6.1.9, it is useful to require t to be well-behaved in the sense of being at least linear; note that almost all nondecreasing functions which are easy to describe are at least linear.

6.1.7 EXERCISE

(a) Let t be such that $t(n) = 2^n$; show that t is at least linear and that for any polynomial p (of one variable) and any positive integer k, $p(n) \leq t(n/k)$ a.e. n.

(b) For each integer $k > 0$, show that the polynomial n^k is at least linear.

(c) Show that the function t such that $t(n) = log_2 n$ is *not* at least linear.

Now, what we wish to prove is that if t is at least linear, if Φ is a linearly bounded complexity measure, and if S is a set rich enough to express the t-limited halting problem for Φ, then any partial decision procedure for S is at least t-difficult (in the measure Φ) on the elements of S. Such a result is really just an extension of a standard version of the halting problem. If the set $T = \{\langle i, x \rangle : \phi_i(x) \text{ is undefined}\}$ were recursive (or even just recursively enumerable), then we could define a total recursive function f such that

$$\phi_{f(i)}(x) = \begin{cases} 1 & \text{if } \langle i, x \rangle \in T \\ divergent & \text{otherwise} \end{cases}$$

Letting i be chosen via the recursion theorem so that $\phi_i = \phi_{f(i)}$, we have that for any x,

$\langle i, x \rangle \in T$ iff $\phi_{f(i)}(x)$ *convergent* iff $\phi_i(x)$ *convergent* iff $\langle i, x \rangle \notin T$.

Notice that under the assumption that the halting problem is (partially) solvable, we have constructed a "self-referencing" program which says, "I shall halt just in case I do not halt."

Now, employing the notation of Definition 6.1.4, we might assume that the set S expresses a t-limited halting problem and that the partial recursive function ϕ_d partially solves S. We could then replace the condition $\langle i, x \rangle \in T$ by the condition $\phi_d(R(i, x))$ *convergent* (that is, $\Phi_i(x) > t(|x|)$), obtaining a partial recursive function ϕ_i via the recursion theorem such that

$$\phi_i(x) = \begin{cases} 1 & \text{if} \quad \phi_d(R(i, x)) \, convergent \\ divergent & \text{otherwise.} \end{cases}$$

That is, $\phi_i(x)$ *convergent* if and only if $R(i,x) \in S$. But then $R(i,x) \notin S$ implies $\Phi_i(x) \le t(|x|)$ by the definition of S, while the definition of i shows that $R(i, x) \notin S$ implies that $\phi_i(x)$ *divergent*. This contradiction shows that $R(i, x) \in S$ for all x. On the other hand, since $R(i, x)$ can be constructed quickly (see Definition 6.1.4), if the application of the recursion theorem to obtain i does not greatly increase the complexity of a calculation, then it is reasonable to hope that $\Phi_i(x)$ will be small whenever $\Phi_d(R(i, x))$ is small. In particular, it is reasonable to expect that $\Phi_i(x) \le t(|x|)$ whenever $\Phi_d(R(i, x))$ is much less than $t(|x|)$. But we already know that $R(i, x) \in S$ for all x, so from the definition of S we can never have $\Phi_i(x) \le t(|x|)$. Thus we can never have $\Phi_d(R(i, x))$ much less than $t(|x|)$.

If you ponder the preceding discussion carefully, you should see that we have once more constructed a "self-referencing" program. This time it says, "If I am going to run quickly, then I shall not halt at all."

The program i whose behavior we have just outlined must be defined recursively, that is, through an application of the recursion theorem. To make the proof work we need to guarantee that this application of the recursion theorem results in at most a linear increase in complexity, that is, that the function h of Theorem 5.4.6 is no worse than a multiplicative constant. Although such a linear bound does hold for most of the complexity measures of Section 1.9 (see Exercise 5.4.14), it does not follow from the definition of a linearly bounded complexity measure that such a bound holds in every such measure, nor does such a linear bound seem to hold in all useful measures, including the MIN*space* measure introduced in Section 2.3 which we use in Section 6.4. The difficulty, at least in the MIN*space* measure, is that the cost of simulation, that is, of computing the universal function, seems nonlinear, and unfortunately simulation in the form of a universal function is used in our proof of the Recursion Theorem (3.4.1). However, as the reader who has worked Exercise 3.4.20 will realize, there is another formulation of the recursion theorem, and it turns out that a direct proof of this other formulation of the recursion theorem can be given without any appeal to a universal function.

6.1.8 RECURSION THEOREM, ANOTHER VERSION *Let* ϕ_0, $\phi_1, \ldots$ *be any acceptable programming system.*

1. *Let* ϕ_j *be any partial recursive function. From j we can effectively find a program n such that for all x*

$$\phi_n(x) = \phi_j(n, x).$$

2. *If* $\Phi_0, \Phi_1, \ldots$ *is any linearly bounded measure on* $\phi_0, \phi_1, \ldots$ *then there is a constant k such that the program n found in part 1 can be chosen to satisfy*

$$\Phi_n(x) \leq k \cdot \Phi_j(n, x) \quad \text{a.e.} \, x.$$

Proof The proof is via a magical rabbit in the hat construction. It may be difficult to see the motivation, but it is nevertheless easy to verify the proof.

Let s be an s-1-n function and c a function for composition with $\phi_{c(i,j)}(x, y) = \phi_i(\phi_j(x), y)$. Let *self* be a fixed program such that $\phi_{self}(y) = s(y, y)$. Then

$$\begin{aligned}
\phi_{s(c(j, self), c(j, self))}(x) &= \phi_{c(j, self)}(c(j, self), x) \\
&= \phi_j(\phi_{self}(c(j, self)), x) \\
&= \phi_j(s(c(j, self), c(j, self)), x).
\end{aligned}$$

Thus if we choose $n = s(c(j, self), c(j, self))$ we have that

$$\phi_n(x) = \phi_j(n, x) \quad \text{for all} \, x.$$

If we now let k_0 be the constant for the linearly bounded measures Φ as described in Definition 6.1.2, then

$$\begin{aligned}
\Phi_n(x) &= \Phi_{s(c(j, self), c(j, self))}(x) \\
&\leq k_0[\Phi_{c(j, self)}(c(j, self), x)] \quad \text{a.e.} \, x \\
&\leq k_0{}^2[\Phi_j(\phi_{self}(c(j, self)), x) + \Phi_{self}(c(j, self))] \quad \text{a.e.} \, x.
\end{aligned}$$

But since $\Phi_j(\phi_{self}(c(j, self)), x) \geq |x|$ and since $\Phi_{self}(c(j, self))$ is a constant independent of x,

$$\Phi_n(x) \leq 2k_0{}^2[\Phi_j(\phi_{self}(c(j, self)), x)] \quad \text{a.e.} \, x,$$

and since $\phi_{self}(c(j, self)) = s(c(j, self), c(j, self)) = n$,

$$\Phi_n(x) \leq 2k_0{}^2[\Phi_j(n, x)] \quad \text{a.e.} \, x.$$

Thus, setting $k = 2k_0{}^2$ proves the theorem. $\qquad \square$

We may turn the argument given after Exercise 6.1.7 into a proof that any set rich enough to express a t-limited halting problem must be t-difficult to decide as follows:

6.1.9 THEOREM *Let* Φ_0, Φ_1, ... *be a linearly bounded complexity measure on the acceptable programming system* ϕ_0, ϕ_1, ... *and let* t *be a total recursive function which is at least linear. Suppose* S *is a set which is rich enough to express the* t-limited halting problem for the measure Φ, and suppose d is any partial decision procedure for S (that is, $\phi_d(y)$ is defined if and only if y is in S). Then there is a (fixed) i such that $R(i,\ x) \in S$ for all x and such that for some positive integer k, $\Phi_d(R(i,\ x)) > t(|R(i,\ x)|/k)$ a.e. x. Moreover, if S is rich enough to strongly express the t-limited halting problem for the measure Φ, then k depends only on the set S and the measure Φ, and not on the program d. (Here, $R(i,\ x) = \phi_r(i,\ x)$ is as in Definition 6.1.4.)*[1]

Proof Suppose that $\phi_d(y)$ is defined if and only if $y \in S$. Then there is a program m such that

$$\phi_m(j,\ x) = \phi_d(R(j,\ x)) = \phi_d(\phi_r(j,\ x)),$$

and by the preceding theorem there is a fixed program i and a constant k' such that

$$\phi_i(x) = \phi_m(i,\ x) \quad \text{for all } x,$$

and

(a) $$\Phi_i(x) \leq k'[\Phi_m(i,\ x)] \text{ a.e. } x.$$

In fact, since $\phi_m(i,\ x) = \phi_{c(d,r)}(i,\ x)$, m is just $c(d,\ r)$ and by Definition 6.1.2 there is a constant k'' such that

(b) $$\Phi_m(i,\ x) \leq k''[\Phi_d(\phi_r(i,\ x)) + \Phi_r(i,\ x)] \text{ a.e. } x.$$

Now if $R(i,\ x) \notin S$, we know that $\Phi_i(x) \leq t(|x|)$, by Definition 6.1.4. Thus $\phi_i(x)$ converges. But $\phi_i(x) = \phi_m(i,\ x) = \phi_d(R(i,\ x))$, which thus also converges, so that $R(i,\ x) \in S$ for all x.

Now choose $k = 2k_ik'k''$ where k_i is chosen as in Definition 6.1.4. Thus

(c) $$|R(i,\ x)| \leq k_i \cdot |x| \quad \text{a.e. } x \text{ and } \Phi_r(i,\ x) < t(|R(i,\ x)|/k) \quad \text{a.e. } x.$$

---

[1] It might seem more natural to state this theorem in terms of *decision procedures* for S (that is, programs for the characteristic function of S) since the theorem is of interest only when S is a recursive set; but in view of Proposition 6.1.5, proving that all *partial* decision procedures for S are complex implies that all decision procedures for S must also be complex. Moreover, although such a development is beyond the scope of this book, *nondeterministic* decision procedures are quite naturally viewed as partial decision procedures, and the theorem as stated is immediately applicable to any precisely formulated theory of nondeterministic computations.

For the sake of a contradiction suppose now that $\Phi_d(R(i, x)) \le t(|R(i, x)|/k)$ where x is chosen large enough to make inequalities (a), (b), and (c) all hold. Then from (a), (b), and (c)

$$\begin{aligned}
\Phi_i(x) &\le k'[\Phi_m(i, x)] \le k'k''[\Phi_d(\phi_r(i, x)) + \Phi_r(i, x)] \\
&\le k'k''[t(|R(i, x)|/k) + t(|R(i, x)|/k)] \\
&\le 2k'k''[t(k_i \cdot |x|/k)] \qquad \text{(since } t \text{ is monotone)} \\
&\le t(2k'k''k_i \cdot |x|/k) \qquad \text{(since } t \text{ is at least linear)} \\
&\le t(|x|).
\end{aligned}$$

But $\Phi_i(x) \le t(|x|)$ means that $R(i, x) \notin S$, contradicting the fact that $R(i, x) \in S$ for all x. We conclude that $R(i, x) \in S$ for all x, but that $\Phi_d(R(i, x)) \le t(|R(i, x)/k)$ for at most finitely many x.

The proof of the theorem is now complete except for the special case when S is rich enough to strongly express the t-limited halting problem. For this case, we note that in the choice of k the constant k_i does not depend on i which depends on d, and hence k is independent of i also, and depends only on the set S and the measure.   □

Theorem 6.1.9 is the tool used in the rest of this chapter to prove lower bounds on the complexities of some specific computational problems. Having it, to show that a problem is intrinsically difficult one "merely" has to show that the problem is rich enough to express an appropriate limited halting problem for some linearly bounded measure. The next two sections illustrate the use of this general method by applying it to examples based on finite automata theory. These two examples were chosen not because we believe them to be extremely important or natural computational problems, but rather because they readily illustrate the use of this general method. Additional examples of intrinsically difficult decision problems are in the literature, and the proofs use somewhat different techniques. Once you understand the next two sections you should be able to apply the general method of this section to obtain somewhat simpler proofs of any of these results.

Additional Exercises

6.1.10 Suppose that in Definition 6.1.2, we drop the requirement that $\Phi_i(x) \ge |x|$, that Condition 1 is replaced by

$$\Phi_{c(i,j)}(\bar{x}, \bar{y}) \le k_{i,j}[\Phi_i(\phi_j(\bar{x}), \bar{y}) + |\bar{y}|\Phi_j(\bar{x})] \quad \text{a.e. } \bar{x}, \bar{y}$$

and

$$\Phi_{c(i,j)}(\bar{x}) \le k_{i,j}[\Phi_i(\phi_j(\bar{x})) + \Phi_j(\bar{x})] \quad \text{a.e. } \bar{x},$$

and that Condition 2 is replaced by $\Phi_{s(i,z)}(\bar{x}) \le k_i[\Phi_i(z, \bar{x})]$ a.e. $\bar{x}$, where $k_{i,j}$ and k_i are constants which depend on i, j and on i, respectively.

(a) Show that TM*time* and the product of Turing machine time and space each satisfy these weaker conditions.

(b) Show that for measures which satisfy this weaker notion of linear boundedness, a slightly weaker version of Theorem 6.1.8 still holds and the part of Theorem 6.1.9 which does not refer to sets which are rich enough to strongly express limited halting problems also holds.

*6.1.11   Show that Φ_0, Φ_1, . . . in Theorems 6.1.8 and 6.1.9 need not actually be a complexity measure in order for the theorem to hold; specifically, verify that we never used the fact that $\Phi_i(x) \leq y$ is a recursive predicate of i, x, and y, but only that domain ϕ_i = domain Φ_i for all i.

6.1.12   Define a linearly bounded measure to be superlinear if it is linearly bounded and in addition there is a universal program u and a "pairing" program p with $\phi_u(i, x) = \phi_i(x)$ and $\phi_{p(x)}(y) = \langle x, y \rangle$ such that for each i and x there are positive constants k_i and k_x with $\Phi_u(i, x) < k_i \cdot \Phi_i(x)$ and $\Phi_{p(x)}(y) < k_x \cdot |y|$; that is, in superlinear measures there is very little overhead for "simulation" and "pairing."

(a) Pick at least two of the complexity measures considered in Section 1.9 and show that they are superlinear.

(b) Show that if Φ_0, Φ_1, . . . is a superlinear measure and f is any total recursive function, then there is a fixed point program n such that $\phi_n = \phi_{f(n)}$ and $\Phi_n(x) \leq k \cdot \Phi_{f(n)}(x)$ a.e. for some positive constant k. *Hint*: carefully examine the proof of the Recursion Theorem (3.4.1).

(c) Give an alternative proof of (the first part of) Theorem 6.1.9 assuming that the measure is superlinear.

6.1.13   This is a good time to go back and do Exercises 3.4.20 and 5.4.14 if you didn't do them earlier.

6.1.14   For any measure Φ and function t, define $S_t = \{\langle i, x \rangle : \Phi_i(x) > t(|x|)\}$. For which of the measures Φ of Section 1.9 and for which functions t can you now use Theorem 6.1.9 to prove that any decision procedure for S_t must be at least t-difficult?

6.2   EXPONENTIAL DIFFICULTY FOR REGULAR EXPRESSIONS WITH SQUARING

In this section we show that the problem of deciding whether or not certain extended regular expressions represent sets with empty complement requires exponential space in all reasonable programming systems; thus this problem, while algorithmically solvable, is intractable in the sense that there is no practical algorithm for solving it. In addition to illustrating the general method of the previous section, the techniques in this section involving regular expressions and Turing machines (including the exercises at the end) are important to an understanding of parts of the next chapter.

We define *regular expressions with squaring* for some fixed alphabet A_k simultaneously with, for each such expression E, the subset $L[E]$ of A_k^* *represented* by E inductively as follows: $\emptyset$ and ϵ are regular expressions with squaring with $L[\emptyset] = \emptyset$ and $L[\epsilon] = \{\epsilon\}$; for each a_i in A_k, a_i is a regular expression with squaring and $L[a_i] = \{a_i\}$; if E and F are regular expressions with squaring, then so are

1. $(E \cup F)$ with $L[(E \cup F)] = L[E] \cup L[F]$,

2. (EF) with $L[(EF)] = L[E]L[F]$,

3. (E^*) with $L[(E^*)] = L[E]^*$, and

4. (E^2) with $L[(E^2)] = L[E]L[E]$.

You should reread Section 1.1 if you have forgotten the definitions of some of these operations on sets of strings. (The reader who is not familiar with regular expressions and feels in need of explanation beyond what we are giving is referred to standard textbooks on finite automata theory.)

6.2.1 EXERCISE   For any finite subset A of A_k^*, show how to write a regular expression with squaring (without using $*$ or squaring) which represents A. What is the length of this expression in terms of A?

Notice that $L[(E^2)] = L[(EE)]$, and so the class of sets represented by regular expressions with squaring is the same as the class of sets represented by "ordinary" regular expressions (that is, expressions without squaring) and in fact one can routinely eliminate all occurrences of squaring to obtain ordinary regular expressions with no change in the sets represented.

6.2.2 EXERCISE   Describe how to carry out this elimination of squaring, and give a bound on the length of the resulting ordinary regular expression in terms of the length of the original regular expression with squaring.

Since the class of sets represented by regular expressions with squaring is the class of regular sets (that is, the class of sets accepted by finite automata), adding squaring to regular expressions does not serve to allow representation of additional sets but rather to provide more succinct representations of some sets.

6.2.3 EXERCISE

(a) Let $E_0 = a_1$ and $E_{n+1} = (E_n{}^2)$ for $n \geq 0$. For each n, what is $L[E_n]$ and what is $|E_n|$?

(b) Let E be a regular expression with squaring such that $L[E]$ contains a word of length m. Define $E_0 = E$ and $E_{n+1} = (E_n{}^2)$ for $n > 0$. For each n, show that $L[E_n]$ contains a word of length $2^n m$. In terms of $|E|$, what is $|E_n|$?

(c) Prove by induction on $|E|$ that if E is an ordinary regular expression with $|E| \leq n$ such that $L[E]$ is a *finite* set then $L[E]$ contains no words of length greater than n.

The problem of deciding for an arbitrary regular expression with squaring E whether $L[E] = A_k{}^*$ is algorithmically solvable. Standard results from finite automata theory suggest a very straightforward algorithm: first convert E to an equivalent ordinary regular expression E'; then convert E' to a nondeterministic finite automaton M such that the set accepted by M, $T(M)$, is $L[E']$; then construct the deterministic subset automaton M' from M (with $T(M) = T(M')$) and interchange accepting and nonaccepting states in M' to get M'' such that $T(M'') = A_k{}^* - L[E]$; and finally test whether $T(M'') = \emptyset$ by seeing whether any accepting state is accessible. This algorithm can be implemented, say, by a Turing machine which will run in space bounded by $2^{2^{c|E|}}$ for some positive integer constant c. However, this algorithm makes very inefficient use of space; to make efficient use of space it should be "reused" as many times as possible. By making clever reuse of space, essentially the same algorithm can be implemented by a Turing machine which runs in space bounded by $2^{c|E|}$ for some positive integer constant c (see Exercise 6.2.9).

Let RE^2 be the set of regular expressions with squaring E such that $L[E]$ has empty complement. We shall show that deciding membership in RE^2 is exponentially difficult by showing that RE^2 is rich enough to express an exponentially limited halting problem for a Turing machine space measure $\Phi_0, \Phi_1, \ldots$. It turns out to be notationally convenient for the proof not to use the Turing machine programming system from Section 1.7, but rather to use a slightly restricted Turing machine programming system instead; we shall use a programming system consisting of Turing machines which never move more than one square to the left of their original starting position on the tape, and which halt only in a single "halting" state h such that there are no instructions beginning "$h \ldots$". From the discussion at the end of Section 1.7, or from Exercise 1.7.7, we know that every Turing machine is equivalent to one which satisfies the first condition and uses essentially the same space. To satisfy the second condition one simply adds an additional state h and instructions $j\ a_m\ a_m\ R\ h$ for each pair $j\ a_m$ such that the machine halts when it is in state j looking at a_m; this change obviously yields an equivalent Turing machine which uses at most one more tape square in any computation. Therefore, there is a Turing machine programming system which meets the two restrictions stated above and this restricted system is an acceptable programming system; moreover, the space measure on this programming system is essentially the same as that on the unrestricted Turing machine programming system. Thus, proving exponential difficulty for RE^2 in this programming system also proves exponential difficulty for RE^2 in the space measure on the unrestricted Turing machine programming system. It is quite easy to show that the space measure on the restricted Turing machine system is linearly bounded; for example, one could redo Exercise 6.1.3.

6.2.4 EXERCISE Verify that the space measure on the restricted Turing machine system is linearly bounded.

In view of Theorem 6.1.8 and Proposition 6.1.5, to show that any Turing machine which decides membership in RE^2 requires exponential space on infinitely many inputs, it is now sufficient to show that RE^2 is rich enough to express the exponentially limited halting problem for the space measure on our restricted Turing machine programming system. It turns out to be notationally simpler to use the function t such that $t(n) = 2^n + n + 2$ rather than 2^n for our application of Theorem 6.1.9.

6.2.5 EXERCISE Show that t defined by $t(n) = 2^n + n + 2$ is at least linear.

Recall from the proof of Theorem 1.8.3 that an instantaneous description of a Turing machine at some point in its computation consists of a string which includes the nonblank portion of the tape with a symbol indicating the state the machine is in inserted directly to the left of the symbol the machine is currently scanning. If Turing machine i has alphabet A and states $0, 1, \ldots, p$ then our alphabet for regular expressions with squaring will be $A_k = A \cup \{r_0, r_1, \ldots, r_p, \#\}$; in this way, expressions can represent sets of strings including strings of the form $\#ID_1\#ID_2\#ID_3\ldots$ where $ID_1, ID_2, ID_3, \ldots$ is a sequence of instantaneous descriptions of i's computation on some input. Since we are interested in whether or not i *halts in space* $t(|x|)$ *on input* x, that is whether $\Phi_i(x) > 2^{|x|} + |x| + 2$, and since our Turing machines only use tape to the right of their starting point, we can assume that *all* instantaneous descriptions have length exactly $2^{|x|} + |x| + 3$ by adding enough blanks to the right-hand end to reach this length (the "3" is to allow for the state symbol). Thus, for input x to Turing machine i, $\Phi_i(x) > t(|x|)$ if and only if there is *no* string $\#ID_1\#ID_2\# \ldots$ describing a computation of i on x which halts, where *each* instantaneous description has length $2^{|x|}+|x|+3$.

A string in A_k^* describes such a computation if it begins correctly, "moves" correctly, and ends correctly; specifically, if

(a) It begins with $\#ID_1\# = \#Br_0xB^{2^{|x|}+1}\#$.

(b) For every triple of symbols $y \in A_k^3$, the triple of symbols y' occurring $2^{|x|}+|x|+1$ spaces to the right (that is, with $2^{|x|}+|x|+1$ intervening symbols) is correct for such a sequence of instantaneous descriptions for i.

(c) It contains a halting configuration for i (that is, it contains the symbol r_h for the halting state h).

Therefore, $\Phi_i(x) > t(|x|)$ if and only if there is *no* word in A_k^* which meets conditions (a) through (c) above. Given i and x, our job now is to show how to write down a regular expression with squaring $R(i, x)$ such that $L[R(i, x)] = A_k^*$ iff $\Phi_i(x) > t(|x|)$, such that $R(i, x)$ has length proportional to $|x|$, and such that the complexity of producing $R(i, x)$ is not too great. We define such an appropriate $R(i, x)$. As is usual mathematical practice, we omit parentheses when they do not matter. We also use some other notational abbreviations to simplify describing $R(i, x)$: if A is a finite subset of A_k^*, we use A itself to stand for a regular expression with squaring which represents A; if A and B are finite sets and w is a word, $A - w$ stands for an expression representing $A-\{w\}$, $A-B$ stands for an expression representing $A-B$, and w stands for an expression representing $\{w\}$. If E is a regular expression with squaring, we let $E^{2^0} = E$ and $E^{2^{n+1}} = ([E^{2^n}]^2)$ for $n \geq 0$; note that $|E^{2^n}|=3n+|E|$, (see Exercise 6.2.3).

It is trivial to represent the set of strings which do not satisfy condition (c) above, that is which do not contain a halting configuration. Let **C** be the expression given by

$$\mathbf{C} = [A_k - r_h]^*;$$

then **C** certainly serves the purpose and $|\mathbf{C}|$ is constant, depending at most on the machine i.

Let the input x be $x = x_1x_2 \ldots x_n$ with each x_j in the alphabet of the Turing machine i. An expression which represents all strings which do not begin correctly, that is which do not satisfy condition (a) above, is given in the two parts $\mathbf{A}'$ and $\mathbf{A}''$ below:

$$\mathbf{A}' = (A_k-\# \cup \#(A_k-B \cup B(A_k-r_0 \\ \cup\, r_0(A_k - x_1 \cup x_1(\ldots \cup x_{n-1}(A_k - x_n) \ldots)))))A_k^*$$

$$\mathbf{A}'' = \#Br_0x(\,[A_k - \#]^* \\ \cup\, [B^*(A_k - \{B, \#\}) \cup \{B, \epsilon\}^{2^n} \cup B^{2^n}B^2B^*]\#A_k^*)$$

A' represents the set of words which do not begin with $\#Br_0 x$; note that to keep $|A'|$ proportional to $|x|$ we had to be a bit tricky. A'' represents the set of words which begin with $\#Br_0 x$ but which do not have a *next occurrence* of $\#$, or which do not have all B's between x and the next $\#$, or do not have enough B's between x and the next $\#$, or have too many B's between x and the next $\#$. Then $A = A' \cup A''$ represents the set of all strings which do not satisfy condition (a), and $|A|$ is proportional to $|x|$.

To give an expression which represents the set of words which do not "move" correctly we define the *move function* M_i from $A_k{}^3$ into the subsets of $A_k{}^3$ as follows: for each instruction $j a_u a_v R m$ in i, $M_i(a_w r_j a_u)$ $= \{a_w a_v r_m\}$ for all a_w in A_k; for each instruction $j a_u a_v L m$ in i, $M_i(a_w r_j a_u)$ $= \{r_m a_w a_v\}$ for all a_w in A_k; for each triple $s_1 s_2 s_3$ in $A_k{}^3$ which *does not* contain any state symbol r_j, $M_i(s_1 s_2 s_3) = A_k s_2 A_k$; and for all other triples y in $A_k{}^3$, $M_i(y) = \emptyset$. Notice that if a string begins with $\#ID\#$ where ID is an instantaneous description of i of length $2^n + n + 3$, and if the string is such that for all triples y the triple y' occurring $2^n + n + 1$ symbols to the right is in $M_i(y)$, then the string gives a sequence of instantaneous descriptions which describe a computation of i in space $2^n + n + 2$. Finally, an expression B which represents the set of all strings which do not satisfy condition (b) above is given by

$$B = A_k^*[\cup_{y \in A_k{}^3, M_i(y) \neq \emptyset} (y A_k^{2^n} A_k^{n+1}[A_k^3 - M_i(y)])]A_k^*.$$

You can think of the expression B as "sweeping" two 3-symbol "windows" which are $2^n + n + 1$ symbols apart along a string and looking for "violations" of the move function M_i. Thus, in constructing B we used the principle, which was also used in the proofs of Theorems 1.8.3 and 2.6.2, that whether a word is a sequence of instantaneous descriptions which correctly describes a Turing machine computation can be checked "locally". Note that $|B|$ is also proportional to $|x| = n$, with the constant of proportionality clearly depending on the Turing machine i.

Therefore $L[A \cup B \cup C]$ is the set of all strings which do not describe a halting computation of i on input x using space no greater than $2^{|x|} + |x| + 2$; that is, $L[A \cup B \cup C] = A_k^*$ if and only if $\Phi_i(x) > t(|x|)$. Thus if RE^2 is the set of regular expressions with squaring which represent sets with empty complements and $R(i, x)$ is $A \cup B \cup C$ we have for each i that $R(i, x)$ is (obviously) one-to-one in x, $R(i, x) \in RE^2$ iff $\Phi_i(x) > t(|x|)$ for all i and x, and for each i there is a positive integer constant k_i such that $|R(i, x)| \leq k_i \cdot |x|$ a.e. x. To conclude that RE^2 is rich enough to express the t-limited halting problem for (restricted) Turing machine space it remains to show that there is some Turing machine r

such that $\phi_r(i, x) = R(i, x)$ for all i and x, and for each i, $\Phi_r(i, x) \le t(|R(i, x)|/k)$ a.e. x for all positive integers k. To show this it is sufficient to show that there is some r such that $\phi_r(i, x) = R(i, x)$ for all i and x, and $\Phi_r(i, x) \le p_i(|x|)$ for all x, where p_i is a polynomial depending on i; see Exercise 6.1.7. It should be intuitively clear that given x and the instructions in a Turing machine i, a simple algorithm could produce $R(i, x)$ in not much more additional space than $|R(i, x)|$, and that this algorithm could easily be implemented by a Turing machine. Assuming we have numbered restricted Turing machines in some reasonable way, similar to our numbering of RAM programs in Section 2.4, so that given i we can effectively (and actually fairly easily) construct the set of instructions in Turing machine i. We are making this assumption.

The alert reader may have noticed by now that we have been a bit sloppy in the following way: for each i, the alphabet A_k for the expressions $R(i, x)$ depends on (the number of states in) i. To get around this inconvenience we shall assume that all expressions over any A_k are coded in some fixed alphabet, say A_3, by using some reasonable coding function, say $C_2 \circ D_k$ (which converts "integers" from base k to base 2), and that the value of this coding function is preceded by, say, $a_1{}^k a_3$ to indicate the value of k. Then for each $k \ge 2$ there will be a positive constant c_k such that the coded version of an expression E over A_k will have length no greater than $c_k \cdot |E|$, and the complexity of dealing with these coded expressions will be very little more than that of dealing with the original expressions. Therefore, we shall henceforth assume that $R(i, x)$ is actually the coded version of $A \cup B \cup C$ and that RE^2 is the set of coded regular expressions with squaring which represent sets with empty complement. Then we still have that $R(i, x) \in RE^2$ iff $\Phi_i(x) > t(|x|)$ for all i and x, and for all i there is a k_i such that $|R(i, x)| \le k_i \cdot |x|$ a.e. x. In addition, we also have that there is an r such that $\phi_r(i, x) = R(i, x)$ for all i and x, and such that $\Phi_r(i, x) \le p_i(|x|)$ for all x, where p_i is a small degree polynomial depending on i.

6.2.6 EXERCISE

(a) Fill in enough of the details missing from the previous two paragraphs to convince yourself of the truth of the last assertion: specifically, show that there is a Turing machine r such that for all i and x, $\phi_r(i, x) = R(i, x)$ and $\text{TM}space_r(i, x) \le k_i(|x| + 1)$ where k_i is a constant depending on i.

(b) Show that in addition, the machine from part (a) can be taken to satisfy $\text{TM}time_r(i, x) \le k_i(|x|^3 + 1)$ for all i and x; this part is not needed for our results in this chapter, but it is used in the next chapter.

Summarizing, we have established that RE^2 is rich enough to express the t-limited halting problem for restricted Turing machine space, that restricted Turing machine space is a linearly bounded complexity measure, and that the function t is at least linear; thus we can apply Proposition 6.1.5 and Theorem 6.1.9. Finally, since restricted Turing machine space is essentially the same as (ordinary) Turing machine space, and since $t(n) > 2^n$ for all n; we have proved the following theorem.

6.2.7 THEOREM *Let d be any Turing machine which decides whether or not regular expressions with squaring represent sets with empty complement. Then there is a positive integer constant k which depends on d such that d uses more than $2^{|E|/k}$ tape squares for infinitely many expressions E (which do represent sets with empty complement).*

Given any natural programming system and a reasonable (space or time) complexity measure on that system, the system can be translated into Turing machine programs in such a way that the space used by the resulting Turing machines will be bounded by a (small degree) polynomial composed with the complexities of the original programs. From this and the previous theorem, it follows that the problem of determining whether regular expressions with squaring represent sets with empty complement is (at least) exponentially difficult in all such measures. Therefore, this computational problem, while algorithmically solvable (by a Turing machine in exponential space, in fact), is intractable in the sense that there is no practical algorithm for solving it.

Additional Exercises

6.2.8 Let d and k be as in Theorem 6.2.7 preceding. Show that $\lim \inf_{n \to \infty} |\{x: |x| \leq n \text{ and } \mathrm{TM} space_d(x) > 2^{|x|/k}\}| / n > 0$, where for any finite set A, $|A|$ stands for the number of elements in A.

*6.2.9 Show how to test whether an ordinary regular expression E represents a set with empty complement in space proportional to $|E|^2$. Specifically, build a nondeterministic finite automaton M such that $T(M) = L[E]$ and M has at most $2|E|$ states. Then test whether there is an accessible nonaccepting state in the subset automaton M' for M by using an *accessibility predicate* AC for M' such that if S_1 and S_2 are sets of states in M' and n is a positive integer, then $AC(S_1, S_2, n)$ if and only if there is some string of length no greater than n which takes M' from state S_1 to state S_2. Note that for all $n \geq 1$, $AC(S_1, S_2, 2n)$ if and only if there is a set of states S_3 in M such that $AC(S_1, S_3, n)$ and $AC(S_3, S_2, n)$. If S_0 is the initial state in M' and S is any nonaccepting state in M', then you want to test whether $AC(S_0, S, 2^{2|E|})$; using the previous observation and reusing space many times you should be able to do this in space proportional to $|E|^2$. (The reader who is not familiar with the implementation of recursive procedures may have to think about this for a little while.)

*6.2.10 For any positive integer c, let the function t_c be defined by $t_c(n) = 2^{cn} + n + 2$ for all n. Show that for any c, the set RE^2 of regular expressions with squaring which represent sets with empty complement is rich enough to express the t_c-limited halting problem for the restricted Turing machine space measure. Specifically, redo Exercise 6.2.6 and the material preceding it with t replaced by t_c. Make sure that you still have a polynomial time bound for obtaining the expression $R(i, x)$.

6.3 SUPEREXPONENTIAL DIFFICULTY FOR LENGTH-EXPRESSIONS

In the previous section we saw that the problem of deciding whether certain extended regular expressions represent sets with empty complement requires exponential space. In this section we show that the problem of deciding, for another class of extended regular expressions, whether the expressions represent the empty set is not elementary recursive. That is, if $ex(0, n) = n$ and $ex(m + 1, n) = 2^{ex(m,n)}$ for all m and n, then for every m, any algorithm which solves this problem has complexity greater than $ex(m, n)$ on some inputs of length n for infinitely many n, in any reasonable complexity measure.

As in the previous section, we define *length-expressions* for the alphabet A_k simultaneously with, for each expression E, the subset $L[E]$ of A_k^* *represented* by E inductively as follows: $\emptyset$ and ϵ are length-expressions with $L[\emptyset] = \emptyset$ and $L[\epsilon] = \{\epsilon\}$; for each a_i in A_k, a_i is a length-expression with $L[a_i] = \{a_i\}$; if E and F are length-expressions then so are

1. $(E \cup F)$ with $L[(E \cup F)] = L[E] \cup L[F]$,

2. $(E \cap F)$ with $L[(E \cap F)] = L[E] \cap L[F]$,

3. (EF) with $L[(EF)] = L[E]L[F]$,

4. (E^*) with $L[(E^*)] = L[E]^*$

5. $(-E)$ with $L[(-E)] = A_k^* - L[E]$, and

6. $L[(\text{leh } E)] = \{x \in A_k^* : |x| = |y| \text{ for some } y \in L[E]\}$.

By standard results of elementary finite automata theory, if E is a length-expression then $L[E]$ is a regular set (showing that the length operator "leh" applied to a regular set yields a regular set is a very easy exercise); thus the addition of $\cap$, $-$, and "leh" to ordinary regular expressions does not add any new sets to the class of sets represented, but only allows more succinct representation of some sets. Moreover, standard results of finite automata theory yield a simple, though sometimes very tedious, algorithm to convert any length-expression into an equivalent ordinary regular expression and then to test whether the set it represents is empty.

6.3.1 EXERCISE Give an algorithm to test whether or not arbitrary length-expressions represent the empty set; if it is reasonably "efficient," your algorithm should run in space proportional to $ex(|E|, |E|)$ on any expression E!

Let S be the set of length-expressions E such that $L[E] = \emptyset$. We shall show that S is not elementary recursive by showing that S is rich enough to express the t_m-limited halting problem for the restricted Turing machine space measure from the previous section, where for each m, $t_m(n) > ex(m, n)$ for all n. In the previous section the key to our proof of exponential difficulty was the fact that by using squaring we could write relatively short expressions which represent very long strings, and thus we could write expressions describing very long sequences of Turing machine instantaneous descriptions (see Exercise 6.2.3 and the definitions of expressions **A** and **B**). In this section we shall follow essentially the same strategy, except that length-expressions give us the power to represent very much longer strings; our first goal is to show how to do this, and we shall do so by using length-expressions to *describe Turing machine computations*.

6.3.2 EXERCISE Let the function f be defined by

$$f(0, n) = (n + 4)[(n + 1)2^{n+1} + 1] + 1$$

and

$$f(m + 1, n) = (f(m, n) + 4)[(f(m, n) + 1)^{f(m,n)+1} + 1] + 1$$

for all m and n. Verify that $ex(m, n) < f(m, n)$ for all m and n.

6.3.3 PROPOSITION *For each m and n there is a length-expression $E(m, n)$ and a word x such that $L[E(m, n)] = \{x\}$ with $|x| = f(m, n)$ and such that there is a positive integer constant k with $|E(m, n)| \leq nk^m$ for all m and n, where f is as in the previous exercise. Moreover, there is a restricted Turing machine which on inputs m and n produces $E(m, n)$ using space no more than $2|E(m, n)|$.*

Proof Let **T** be the restricted Turing machine with state transition diagram given in Figure 6.3.4; note that on input 0^n, **T** "counts" to 2^n. That is, on input 0^n, **T** runs for time (exactly) $(n+1)2^{n+1}$ in space $n+2$ (the space originally occupied by $B0^nB$). Thus if $N = (n+1)2^{n+1}+1$ and $x = \#ID_1\#ID_2\# \ldots \#ID_N\#$ is the sequence of instantaneous descriptions each of length $n+3$ for **T**'s computation on input 0^n, then $|x| = (n+4)N+1 = (n+4)[(n+1)2^{n+1}+1]+1$.

For writing length-expressions in this section, we shall use the same abbreviations used for writing regular expressions with squaring in the previous section. Note that the alphabet for length-expressions to describe **T**'s computations is $A = A_8 = \{B, 0, 1, r_0, r_1, r_2, r_3, \#\}$ with $r_3 = r_h$, and let the move function M_T for **T** be defined as in the previous section (for the definition of the expression **B** given there). For any length-expression E, consider the length-expression $E_T[E] = A \cap B \cap C \cap D$ where **A**, **B**, **C**, and **D** are given below:

$$\mathbf{A} = \#Br_0[0^* \cap \text{leh}E]B\#A^*$$
$$\mathbf{B} = -(A^*[\cup_{y \in A^3, M_T(y) \neq \emptyset} (y(\text{leh}EB)[A^3 - M_T(y)])]A^*)$$
$$\mathbf{C} = A^*r_hA^*$$
$$\mathbf{D} = A^*\#.$$

Suppose that $L[E] = \{y\}$ with $|y| = n$; then $L[E_T[E]] = \{x\}$ where x is the sequence of instantaneous descriptions each of length $n+3$ for **T**'s computation on input 0^n. That is, expression **A** represents the set of strings which begin with $\#ID_1\# = \#Br_00^nB\#$; expression **B** represents the set of strings which "move" correctly (for each triple y the triple y' of symbols occurring $n+1$ symbols to the right are correct); **C** represents the set of all strings containing the halting configuration for **T**; and since by the definition of M_T expression **B** allows the last symbol in the string to be anything (not necessarily a $\#$), we have added expression **D** to remedy this. Note that there is a positive integer constant k such that for all E, $|E_T[E]| < k|E|$; since we know what the Turing machine **T** is we

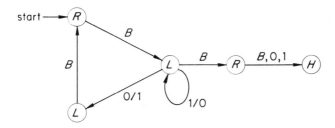

FIGURE 6.3.4 The "counting" Turing machine T.

could actually calculate k, but why bother? Also, there is obviously a restricted Turing machine which on input E produces $E_T[E]$ using space no more than $|E| + |E_T[E]|$.

Now define $E(0, n) = E_T[0^n]$ and $E(m + 1, n) = E_T[E(m, n)]$ for all m and n. Then for all m and n, $L[E(m, n)] = \{x\}$ with $|x| = f(m, n)$ and $|E(m, n)| \le nk^m$. Also, clearly there is a restricted Turing machine which on input m, n produces $E(m, n)$ using space no more than $2|E(m, n)|$, which completes the proof of the proposition. $\square$

With the length-expressions provided by the previous proposition, we can now easily prove that S is rich enough to express limited halting problems for *very* large limits. We shall assume, as we did in the previous section, that all length-expressions over any A_k are coded into some fixed alphabet by using some reasonable coding scheme; see the discussion in the paragraph preceding Exercise 6.2.6. For all m and n, let $t_m(n) = f(m, n)+n+1$ where f is from Exercise 6.3.2. Note that $t_m(n) > f(m, n) > ex(m, n)$.

6.3.5 EXERCISE   Show that for each m, t_m is at least linear.

6.3.6 PROPOSITION   *Let S be the set of (coded) length-expressions E such that $L[E] = \emptyset$. For each m, S is rich enough to express the t_m-limited halting problem for the restricted Turing machine space measure.*

Proof   Let m be fixed. Given a restricted Turing machine i with alphabet A_k, let M_i be the move function for i as defined in the previous section. Define $R(i, x)$ for x in A_k^* to be the (coded) length-expression $\mathbf{A} \cap \mathbf{B} \cap \mathbf{C}$ with $\mathbf{A}$, $\mathbf{B}$, and $\mathbf{C}$ given below:

$$\mathbf{A} = \# Br_0 x[B^* \cap \text{leh}E(m, |x|)]\# A_k^*$$
$$\mathbf{B} = -(A_k^*[\cup_{y \in A_k^3, M_i(y) \ne \emptyset} (y(\text{leh}E(m, |x|))[A_k^3 - M_i(y)])]A_k^*)$$
$$\mathbf{C} = A_k^* r_h A_k^*.$$

$R(i, x)$ represents the set of all strings which begin correctly, "move" correctly, and end correctly to describe a halting computation of machine i on input x in space no more than $f(m, |x|) + |x| + 1 = t_m(|x|)$. Thus, $L[R(i, x)] = \emptyset$ if and only if $\Phi_i(x) > t_m(|x|)$, where Φ_0, Φ_1. . . is the restricted Turing machine space measure. Also, there is a k_i such that $|R(i, x)| \le k_i|x|$ and there is a restricted Turing machine r such that $\phi_r(i, x) = R(i, x)$ and $\Phi_r(i, x) \le k_i(|x|+1)$. Therefore, S is rich enough to express the t_m-limited halting problem for the restricted Turing machine space measure. $\square$

Combining the previous proposition and exercise, Proposition 6.1.5,

Theorem 6.1.9, and Exercise 6.3.2 with the results of Section 1.10 we have a proof of the following theorem:

6.3.7 THEOREM *The problem of deciding whether or not an arbitrary length-expression represents the empty set is not elementary recursive.*

Additional Exercises

6.3.8 Let d be a restricted Turing machine which decides whether length-expressions represent the empty set, and let k be given by Theorem 6.1.9 with $t = t_m$. Show that

$$\lim \inf_{n \to \infty} |\{x: |x| \le n \text{ and } \Phi_d(x) > t_m(|x|/k)\}| / n > 0,$$

where if A is a finite set, $|A|$ denotes the number of elements in A.

6.3.9 Let k be the constant from Proposition 6.3.3, and define $g(n) = f(\log_k n, 0) + n + 1$ for all n.

  (a) Show that for all $c \ge k$, $c \cdot g(n) \le g(c \cdot n)$ a.e. n, and that for all m, $ex(m, n) < g(n)$ a.e. n.

  (b) Show that S (as discussed in this section) is rich enough to express the g-limited halting problem for the restricted Turing machine space measure.

  (c) Show that if d is any restricted Turing machine which decides membership in S, then there is some positive integer constant c and there are infinitely many length-expressions E which represent the empty set such that $\Phi_d(E) > g(|E|/c)$.

6.4 EXPONENTIAL DIFFICULTY OF THE THEORY OF ADDITION

In Chapter 4 we proved the Gödel Undecidability Theorem for Truth, which states that there is no algorithm for deciding which sentences in the language of arithmetic, $\mathscr{L}$, are true in the usual interpretation of the natural numbers under addition and multiplication. The basis of our proof was the representability of all min-computable functions from the very minimal set of axioms given in Definition 4.1.4, and the General Undecidability Theorem shows that any consistent mathematical theory powerful enough to prove this minimal fragment of arithmetic will also be undecidable. Given this, it is natural to ask just how weak a mathematical theory must be in order to be decidable. It turns out that there are some significant fragments of the theory of arithmetic which *are* decidable, but unfortunately they are not "practically" decidable. In this optional section and the next one, we consider such a decidable fragment of arithmetic.

Let the *language of addition* be the language of arithmetic, $\mathscr{L}$, with the function symbol $\otimes$ removed, that is, all the formulas of $\mathscr{L}$ which do not contain the symbol $\otimes$, and let the *theory of addition* (often called *Presburger arithmetic*) be the set of sentences of the language of addition which are true in the usual interpretation of the natural numbers with $\oplus$ interpreted as addition, **0** interpreted as 0, and **1** interpreted as 1. For the remainder of this chapter, whenever we refer to a sentence in the language of addition as being true, we mean true in the usual interpretation.

In this section we show that the theory of addition is exponentially difficult to decide for Turing machine time, and hence is intractable in the sense that there are no practical algorithms for deciding whether a sentence in the language of addition is true. In the exercises at the end, we outline a proof that the theory of addition has doubly exponential difficulty. Finally, in the last section of this chapter we give a decision procedure for the theory of addition, since it is by no means obvious that this theory is actually decidable.

We would like to prove that the theory of addition is rich enough to express an exponentially limited halting problem for a natural complexity measure such as Turing machine time, and thereby conclude that any Turing machine (and hence any program in any natural programming system) must require exponential time to decide truth for sentences in the language of addition. In the previous two sections, the classes of extended regular expressions which we considered quite naturally expressed limited halting problems for Turing machine space. However, getting sentences in the language of addition to "talk about" Turing machine computations is not nearly so simple and natural. Since the general method developed in Section 6.1 can be applied to any halfway reasonable complexity measure on any halfway reasonable programming system, it is more straightforward to apply that general method to a suitably chosen programming system and complexity measure to get an intrinsic difficulty result for that measure, and then to relate that complexity measure to other, perhaps more natural, complexity measures. This is the course we shall pursue.

We begin with the observation that if multiplication could be defined in the language of addition, then our results in Chapter 4 would show the theory of addition to be undecidable, which it is not. Obviously, using sufficiently long formulas in the language of addition we could describe any finite portion of the "multiplication table." Moreover, your intuitions about limited halting problems should strongly suggest that if we could easily write short formulas in the language of addition which describe large (possibly infinite) portions of the multiplication table, then we might be able to show that the theory of addition is rich enough to express *some* suitably large limited halting problem. Obtaining such formulas for multiplication is our first goal.

To get our short formulas for multiplications, we shall need to use two tricks for transforming formulas which obviously "say" what we want but are too long, into logically equivalent, but shorter, formulas. We now illustrate these tricks with examples which are notationally simpler than the use we shall soon make of them. Let $F(z)$ be a formula with the free variable z and let $G(u,v,w,x,y)$ be a formula with the free variables u, v, w, x, and y. Suppose we are interested in "saying" of y,

$$\exists u \, \exists v \, \exists w \, \exists x \, [F(u) \,\&\, F(v) \,\&\, F(w) \,\&\, F(x) \,\&\, G(u, v, w, x, y)].$$

If F is very long, then the length of this formula is excessively long because of the four occurrences of F in it; if we could get down to one occurrence of F we might have a considerably shorter formula. Consider

$$\exists u \, \exists v \, \exists w \, \exists x \, \forall t \, [\, ((u=t \lor v=t \lor w=t \lor x=t) \Rightarrow F(t))$$
$$\&\, G(u, v, w, x, y)],$$

this formula "says" exactly the same thing of y (that is, it is logically equivalent to the formula above), and it uses the (very long) formula F only once.

For the second trick, notice that in order to have an infinite supply of variables we would need to subscript them (say with integers to some base $k \geq 2$), and a proper accounting of the length of formulas would have to include the lengths of these subscripts. Thus, if we are really interested in writing (infinitely many) formulas which are as short as possible, we would like to do so with a finite set of variables. As we pointed out in Section 4.1, there is nothing to prevent a variable from having both bound and free occurrences in the same formula, although this is generally considered bad form because it makes formulas difficult to read. However, in the interest of getting short formulas we shall use this bad form and "re-use" variables in the following way: let formulas $F(u,v,w)$ and $G(x,y,z)$ have free variables u, v, w, and x, y, z, respectively; suppose we wish to "say" of w

$$\exists u \, \forall v \, [F(u, v, w) \,\&\, \exists x \, \exists y \, \forall z \, G(x, y, z)].$$

If we want to be very stingy with variables, it is logically equivalent (although more difficult to read) to "say"

$$\exists u \, \forall v \, [F(u, v, w) \,\&\, \exists u \, \exists v \, \forall w \, G(u, v, w)].$$

This is the other trick we will use to get short formulas.

6.4.1 PROPOSITION   *For each natural number k there is a formula $M_k(x_1, x_2, x_3)$ in the language of addition which has free variables x_1, x_2, and x_3 (and eleven bound variables) such that for any natural numbers m, n, and p, the sentence $M_k(\mathbf{m}, \mathbf{n}, \mathbf{p})$ is true if and only if $m \cdot n = p$ and $m < 2^{2^{2k}}$; moreover, there is a positive integer c such that $|M_k| \leq c(k + 1)$.*

Proof   Let $M_0(x_1, x_2, x_3)$ be the formula

$$(x_1 = 0 \ \& \ x_3 = 0) \bigvee (x_1 = 1 \ \& \ x_2 = x_3).$$

Clearly this has the desired properties. We now proceed inductively, assuming that we have $M_k(x_1, x_2, x_3)$ with the desired properties. Note that $2^{2^{2k+1}} = (2^{2^{2k}})^2$. Using much the same idea as in the outlined proof of "global speedup" for RAM pseudospace in Proposition 5.1.7, we observe that for all n, $(n+1)^2 - 1 = n^2 + 2n$ so that every number less than $(n+1)^2$ can be expressed (although not uniquely) as a sum $x^2 + y + z$ where x, y, and z are less than $n + 1$. In addition, if x, y, and z are less than $n + 1$ then $x^2 + y + z$ is less than $(n + 1)^2$. Therefore, $x_1 < 2^{2^{2k+1}}$ and $x_1 \cdot x_2 = x_3$ if and only if there are u_1, u_2, and u_3, each less than $2^{2^{2k}}$ such that

$$x_1 = u_1^2 + u_2 + u_3 \quad \text{and} \quad x_3 = u_1 \cdot (u_1 \cdot x_2) + u_2 \cdot x_2 + u_3 \cdot x_2.$$

(Notice that this is really just a slight variation on the usual multiplication algorithm which obtains products as the sum of products of smaller numbers.)

Assuming we have a formula M_k describing limited multiplications as defined in the proposition, we can use it inductively to write a formula for multiplications $x \cdot y = z$ where $x < 2^{2^{2k+1}}$. For notational simplicity, if $v_1, \ldots, v_n$ is an n-tuple of variables we shall abbreviate it with $\bar{v}$, and, for example, $\exists \bar{v}$ will abbreviate the sequence of quantifiers $\exists v_1 \exists v_2 \cdots \exists v_n$. Now the formula $M(\bar{x})$ below describes such multiplications:

$$\exists \bar{u} \ \exists s \ \exists \bar{p}(M_k(u_1, u_1, s) \ \& \ x_1 = (s \oplus (u_2 \oplus u_3)) \ \&$$
$$M_k(u_1, x_2, p_1) \ \& \ M_k(u_1, p_1, p_2) \ \& \ M_k(u_2, x_2, p_3) \ \&$$
$$M_k(u_3, x_2, p_4) \ \& \ x_3 = (p_2 \oplus (p_3 \oplus p_4)) \ ).$$

Using our assumptions for M_k and the discussion in the previous paragraph, it is easy, although a bit tedious, to verify that M does indeed describe such multiplications. However, since M has five occurrences of M_k, it is much too long; if we used this construction twice to obtain M_{k+1} we would have $|M_{k+1}| \geq 25|M_k|$, which would yield an unacceptable exponential growth in the length of M_k as a function of k. So we use the first trick above to get the shorter equivalent formula $M'(\vec{x})$ below:

$$\exists \vec{u} \, \exists s \, \exists \vec{p} \, \forall \vec{x}'$$
$$(x_1 = (s \oplus (u_2 \oplus u_3)) \, \& \, x_3 = (p_2 \oplus (p_3 \oplus p_4)) \, \&$$
$$(((u_1 = x_1' \, \& \, u_1 = x_2' \, \& \, s = x_3') \lor (u_1 = x_1' \, \& \, x_2 = x_2' \, \& \, p_1 = x_3') \lor$$
$$(u_1 = x_1' \, \& \, p_1 = x_2' \, \& \, p_2 = x_3') \lor (u_2 = x_1' \, \& \, x_2 = x_2' \, \& \, p_3 = x_3') \lor$$
$$(u_3 = x_1' \, \& \, x_2 = x_2' \, \& \, p_4 = x_3')) \Rightarrow M_k(\vec{x}'))).$$

where $\vec{x}'$ stands for x_1', x_2', x_3'. (We warned you the examples were a lot simpler!) Now let $MESS(\vec{x}, \vec{x}')$ stand for the string of symbols between $\forall \vec{x}'$ and $\Rightarrow$ in the formula $M'(\vec{x})$. We must be careful to remember that $MESS$ is not a formula, but nevertheless we can use it to abbreviate $M'(\vec{x})$ as follows:

$$\exists \vec{u} \, \exists s \, \exists \vec{p} \, \forall \vec{x}' \, MESS(\vec{x}, \vec{x}') \Rightarrow M_k(\vec{x}'))).$$

If $MESS(\vec{x}', \vec{x})$ stands for the string $MESS(\vec{x}, \vec{x}')$ with each occurrence of x_1, x_2, or x_3 replaced x_1', x_2', or x_3' and each occurrence of x_1', x_2', x_3' replaced by x_1, x_2, or x_3, respectively, then the formula $M'(\vec{x}')$ may be abbreviated as follows:

$$\exists \vec{u} \, \exists s \, \exists \vec{p} \, \forall \vec{x} \, MESS(\vec{x}', \vec{x}) \Rightarrow M_k(\vec{x}))).$$

We have just switched $\vec{x}$ and $\vec{x}'$ in M'.

Now we use the second trick above and let $M_{k+1}(\vec{x})$ be the formula

$$\exists \vec{u} \, \exists s \, \exists \vec{p} \, \forall \vec{x}' \, MESS(\vec{x}, \vec{x}') \Rightarrow M'(\vec{x}'))),$$

which in less abbreviated form is

$$\exists \vec{u} \, \exists s \, \exists \vec{p} \, \forall \vec{x}' \, MESS(\vec{x}, \vec{x}') \Rightarrow \exists \vec{u} \, \exists s \, \exists \vec{p} \, \forall \vec{x} \, MESS(\vec{x}', \vec{x}) \Rightarrow M_k(\vec{x})))).$$

By expanding the abbreviations, it is clear that x_1, x_2, and x_3 are the free variables in M_{k+1} and $\vec{u}, s, \vec{p}$, and $\vec{x}'$ are the (eleven) bound variables in M_{k+1}; moreover, assuming that M_k satisfies the conditions of the proposition, then for any natural numbers m, n, and p, the sentence $M_{k+1}(\mathbf{m}, \mathbf{n}, \mathbf{p})$ is true if and only if $m \cdot n = p$ and $m < 2^{2^{2^{k+2}}}$. Finally, note that if X is the *string*

$$\exists \vec{u} \, \exists s \, \exists \vec{p} \, \forall \vec{x}' \, MESS(\vec{x}, \vec{x}') \Rightarrow \exists \vec{u} \, \exists s \, \exists \vec{p} \, \forall \vec{x} \, MESS(\vec{x}', \vec{x}) \Rightarrow$$

and P is the *string*)))) then (by a simple induction on k) the *formula* $M_k(\vec{x})$ is the *string* $X^k M_0(\vec{x}) P^k$ for all natural numbers k. Thus, $|M_k(\vec{x})| = k(|X| + |P|) + |M_0(\vec{x})|$, and so there is a positive integer c such that $|M_k(\vec{x})| \leq c(k + 1)$ for all k. □

Since the proof of the previous proposition shows that the formulas $M_k(\vec{x})$ actually have an extremely simple *form* as a string of symbols, it is clear that given k it is very easy to write down $M_k(\vec{x})$:

6.4.2 EXERCISE Carefully describe a Turing machine which, for some constant d, on any input $k \geq 0$ produces M_k in time no greater than $d \cdot |M_k| \cdot |k|$. It should run in space $|M_k|$ as well.

Now to see how the proof of the General Undecidability Theorem, and in particular Theorem 4.2.6, can be changed to yield an application of Theorem 6.1.9, we first recall the definition of MIN-programs (see Exercise 2.3.7). If $\mathbf{P}$ is a MIN-program of n arguments, $\mathbf{P}(x_1, \ldots, x_n)$ will stand for the partial function it computes. $\oplus$, $\otimes$, and $\mathbf{c}_=$ are MIN-programs of two arguments, with $\oplus(x,y) = x + y$, $\otimes(x,y) = x \cdot y$, and $\mathbf{c}_=(x,y)$ the characteristic function of the equality predicate, respectively. For $1 \leq j \leq n$, $\mathbf{P}_j^n$ is a MIN-program of n arguments with $\mathbf{P}_j^n(x_1, \ldots, x_n) = x_j$. If $\mathbf{H}$ is a MIN-program of m arguments and $\mathbf{G}_1, \ldots, \mathbf{G}_m$ are MIN-programs each of n arguments then $\mathbf{H}(\mathbf{G}_1, \ldots, \mathbf{G}_m)$ is a MIN-program of n arguments with

$$\mathbf{H}(\mathbf{G}_1, \ldots, \mathbf{G}_m)(\vec{x}) = \mathbf{H}(\mathbf{G}_1(\vec{x}), \ldots, \mathbf{G}_m(\vec{x})).$$

Finally, if $\mathbf{H}$ is a MIN-program of $n+1$ arguments then **min z H** is a MIN-program of n arguments with

$$\mathbf{min\ z\ H}(\vec{x}) = min\ z\ [\mathbf{H}(\vec{x},z) = 0].$$

Notice that this is just a straightforward definition of MIN-programs.

To prove Theorem 4.2.6, we gave a translation S of these programs into formulas of $\mathscr{L}$ as follows:

$$
\begin{aligned}
S(\oplus) \quad &\text{is} \quad x \oplus y = z \\
S(\otimes) \quad &\text{is} \quad x \otimes y = z \\
S(\mathbf{P}_j^n) \quad &\text{is} \quad x_j = z \\
S(\mathbf{c}_=) \quad &\text{is} \quad (x = y \ \& \ z = \mathbf{1}) \bigvee (x \neq y \ \& \ z = \mathbf{0}).
\end{aligned}
$$

Similarly, $S(\mathbf{H}(\mathbf{G}_1, \ldots, \mathbf{G}_m))$ is

$$\exists y_1 \ldots \exists y_m [S(\mathbf{G}_1)(\vec{x},y_1) \ \& \ \ldots \ \& \ S(\mathbf{G}_m)(\vec{x},y_m) \ \& \ S(\mathbf{H})(\vec{y},z)]$$

and $S(\mathbf{min\ z\ H})$ is

$$\forall w[w \leq z \Rightarrow \exists y[S(\mathbf{H})(\vec{x},w,y) \ \& \ (y = \mathbf{0} \Leftrightarrow w = z)]].$$

We know from Section 4.2 that for any program $\mathbf{P}$ and integers $\vec{m}$ and q that $\mathbf{P}(\vec{m}) = q$ if and only if $S(\mathbf{P})(\vec{m},q)$ is true (in fact provable from the axioms of Chapter 4, although this is not relevant here).

If in the formulas produced by the translation S we replace all occurrences of the formula representing multiplication by M_k (or a slight variation of M_k) then we would have formulas in the language of *addition* which "almost" represent the functions computed by the programs. Specifically, define a new translation T so that for all k

$$T(\otimes,k) \quad \text{is} \quad M_k(x,y,z) \,\&\, M_k(y,x,z)$$

and otherwise the definition of $T(\mathbf{P},k)$ reads the same as the definition of $S(\mathbf{P})$. Then for any MIN-program $\mathbf{P}$ and integers $\bar{m}$ and q, $T(\mathbf{P},k)(\bar{\mathbf{m}},\mathbf{q})$ is true if and only if $\mathbf{P}(\bar{m}) = q$ and in $\mathbf{P}$'s computation of q all numbers multiplied are less than $2^{2^{2^k}}$.

6.4.3 EXERCISE Verify the assertion we have just made.

To apply Theorem 6.1.9, we need an appropriate complexity measure on MIN-programs (which of course use integers in unary, so $|x| = x$). The preceding paragraph suggests the following measure: if $\mathbf{P}$ is a MIN-program then $\Phi_{\mathbf{P}}(\bar{x}) \le y$ if and only if $\mathbf{P}$ on input $\bar{x}$ is convergent and in every multiplication used by $\mathbf{P}$ in computing $\mathbf{P}(\bar{x})$ both of the integers multiplied are no greater than y. Formally,

$$\Phi_{\mathbf{P}}(\bar{x}) = max\{\bar{x}, max\{a,b: \mathbf{P} \text{ multiplies } a \text{ times } b \text{ in computing } \mathbf{P}(\bar{x})\}\},$$

provided $\mathbf{P}(\bar{x})$ is convergent. It is not clear that this even defines a complexity measure; *if* we know that $\mathbf{P}(\bar{x})$ is convergent, then we can run the computation of $\mathbf{P}(\bar{x})$ and find $\Phi_{\mathbf{P}}(\bar{x})$ as defined above. Although it follows from results in this section and in the last section of this chapter that Φ as just defined is in fact a complexity measure, by Exercise 6.1.11 one does not need a complexity measure in order to apply Theorem 6.1.9. It is sufficient to know that Φ meets most of the requirements for being a linearly bounded complexity measure.

6.4.4 EXERCISE For each constant function, give a MIN-program for computing it which performs no multiplications (see the proof of Proposition 2.3.1). Then verify that Φ satisfies the definition of a linearly bounded complexity measure with the possible exception of the requirement that $\Phi_{\mathbf{P}}(\bar{x}) \le y$ be a recursive predicate of $\mathbf{P}$, $\bar{x}$, and y (see Exercise 6.1.3). How does Φ relate to the MIN*space* measure of Exercise 2.3.7?

The paragraph preceding Exercise 6.4.3 can now be restated as the following proposition.

6.4.5 PROPOSITION *There is an effective translation which, given any MIN-program* $\mathbf{P}$ *of n arguments and any* $k \ge 0$, *produces a formula* $T(\mathbf{P},k)$ *in the language of addition with free variables among* $x_1, \ldots, x_n, y$ *such that for any natural numbers* $m_1, \ldots, m_n$, p *the sentence* $T(\mathbf{P}, k)(\mathbf{m}_1, \ldots, \mathbf{m}_n, \mathbf{p})$ *is true if and only if* $\mathbf{P}(m_1, \ldots, m_n) = p$ *and* $\Phi_{\mathbf{P}}(m_1, \ldots, m_n) < 2^{2^{2^k}}$. *Moreover, for some positive integer constant d depending on* $\mathbf{P}$, $|T(\mathbf{P}, k)| \le d \cdot |M_k|$.

6.4.6 EXERCISE   Verify the last sentence in the previous proposition and show that the translation is not only effective, but that it is in fact very easy to compute; specifically, show that there is a constant b and a Turing machine which on input $\mathbf{P}$, k produces $T(\mathbf{P}, k)$ in time no greater than $b \cdot |T(\mathbf{P}, k)|^3$. $\square$

Let $t(k) = 2^{2^{2k}} - 1$; clearly t is at least linear. Thus to apply Theorem 6.1.9 and get an intrinsic difficulty result for the theory of addition in (the complexity measure) Φ we need only show that this theory is rich enough to express the t-limited halting problem for Φ. The translation of Proposition 6.4.5, together with the preceding exercise, comes very close to establishing this.

6.4.7 PROPOSITION   *The theory of addition is rich enough to express the t-limited halting problem for (the complexity measure) Φ.*

Proof   Let TA be the set of sentences in the language of addition which are true in the usual interpretation. For each MIN-program $\mathbf{P}$ of one argument and each $k \geq 0$ let $R(\mathbf{P}, k)$ be the sentence $\sim\exists\, x\, \exists\, y[T(\mathbf{P}, k)(x, y)$ & $x = \mathbf{k}]$, where $T(\mathbf{P}, k)$ is from Proposition 6.4.5. Recall that MIN-programs use the natural numbers represented in unary and so $|k| = k$ and $|\mathbf{k}| \leq 5k$ (see the discussion just before Exercise 4.1.3) for all k. By Proposition 6.4.5, $R(\mathbf{P}, k) \in TA$ iff $\Phi_P(k) > t(|k|)$, and by Propositions 6.4.1 and 6.4.5 there is a positive integer constant a such that $|R(\mathbf{P}, k)| \leq a \cdot k$ for $k > 0$. It remains to check that Condition 3 of Definition 6.1.4 is satisfied; that is, that $R(\mathbf{P}, k)$ can be produced fairly "easily" in (the measure) Φ.

By Exercise 6.4.6 there is a Turing machine which on inputs $\mathbf{P}$ and k produces $R(\mathbf{P}, k)$ in time proportional to $|R(\mathbf{P}, k)|^3$. From Theorem 2.3.10 it follows that there is a MIN-program $\mathbf{R}$ and a constant c such that $\mathbf{R}(\mathbf{P}, k) = R(\mathbf{P}, k)$ and $\Phi_R(\mathbf{P}, k) \leq 2^{c|R(\mathbf{P},k)|^2}$ since in fact $\mathbf{R}$ never uses any numbers larger than $2^{c|R(\mathbf{P},k)|^2}$ in its computation on inputs $\mathbf{P}$ and k. Finally, since t is doubly exponential this suffices to establish Condition 3 of Definition 6.1.4.

6.4.8 EXERCISE   Verify the assertion we have just made. $\square$

Applying Theorem 6.1.9, we see that the theory of addition has doubly exponential difficulty in (the measure) Φ. To relate this difficulty to natural complexity measures, we use the results in Section 1.9 and Theorem 2.3.10; suppose that some Turing machine decided membership in TA in time exponential in the length of its inputs. Then this Turing machine would translate to a MIN-program which decided membership in TA using numbers in its computations no more than doubly exponential in the value (equal to the length) of *its* inputs.

Combining this with Theorem 6.1.9 (and Proposition 6.1.5) and Propositions 6.4.5 and 6.4.7, we have a proof of the following theorem.

6.4.9 THEOREM *For any Turing machine* **d** *which gives a (partial) decision procedure for the theory of addition there is a constant k and infinitely many true sentences S in the language of addition such that* $TMtime_d(S) > 2^{|S|/k}$. *Thus, the theory of addition has no "practical" decision procedures.*

6.4.10 EXERCISE Fill in the details for the proof of the theorem above.

Additional Exercises

6.4.11 Show how to replace the multiple occurrences of the formula M_k in the sentence $R(\mathbf{P}, k)$ by a single occurrence. Use this to show that there is a constant c such that $|R(\mathbf{P}, k)| \le c(|\mathbf{P}| + |M_k| + |k|)$. Then, conclude that the theory of addition is rich enough to *strongly* express the t-limited halting problem for Φ. Finally, conclude that in Theorem 6.4.9 the constant k can be taken to be independent of the Turing machine **d**.

6.4.12 Define a *language of bounded arithmetic* $\mathscr{LBA}$ by adding a new function symbol **f** to the language of arithmetic $\mathscr{L}$ and making the following changes in the definition of $\mathscr{L}$: if t is a term then $\mathbf{f}(t)$ is also a term; if F is a formula and x is a variable then $\exists\, x \le t\, F$ and $\forall\, x \le t\, F$ are also formulas for any term t which *does not* contain x, and $\exists\, x\, F$ and $\forall\, x\, F$ are *not* formulas. Otherwise the definition of $\mathscr{LBA}$ reads the same as that of $\mathscr{L}$. Let **f** be interpreted as some fixed function $f: N \to N$, and let the theory TBA_f be the set of sentences which are true in the usual interpretation.

(a) Show that the MIN*space* measure is linearly bounded.
(b) Prove that if f is at least linear and larger than exponential then TBA_f is rich enough to express the f-limited halting problem for MIN*space*.
(c) Show that if $f(x) = ex(x,x)$ where ex is from Section 1.10 then the theory TBA_f is not elementary.
(d) Prove that if f is *any* total recursive function then the theory TBA_f is decidable. *Caution:* Even if f is primitive recursive, this does not follow *directly* from Exercise 1.4.13.

The remaining exercises of this section may be used by the truly dedicated reader who wishes to prove that the theory of addition is even harder (by at least another exponential factor) than Proposition 6.4.7 and Theorem 6.4.9 indicate. To improve the lower complexity bounds of Proposition 6.4.7 and Theorem 6.4.9, one needs to know a little more logic and more number theory than fall within the scope of this book. In particular, one needs some information on the distribution, or density, of the prime numbers in the natural numbers. For each n in N, let *primes*(n) be the number of primes no greater than n. A fundamental result on the "size" of the function *primes* is the Prime Number Theorem which states that *primes*(n) is asymptotically equal to $n/ln\ n$, where $ln\ n$ stands for the natural logarithm of n. Unfortunately, all of the known proofs of the Prime Number Theorem are quite long and complicated. However, we do not need the full strength of the Prime Number Theorem for our purposes. Chebychev's Theorem, an approximation to the Prime Number Theorem, which states that for all $n \geq 2$,

$$n/4log_2 n < primes(n) < 32n/log_2 n,$$

is strong enough. For a proof of this result you may consult the book on number theory listed in the section on References and Further Readings at the end of this book.

6.4.13 Let $p_1, \ldots, p_n$ be distinct prime numbers, and let q be their product. Prove that for any $x_1, \ldots, x_n$ such that $x_i < p_i$ for each $1 \leq i \leq n$, there is a unique $y < q$ such that $y\ mod\ p_i = x_i$ for each $1 \leq i \leq n$; i.e., let $f: \{0,1, \ldots, q-1\} \rightarrow \{0,1, \ldots, p_1 - 1\} \times \cdots \times \{0,1, \ldots, p_n - 1\}$ be defined by $f(y) = (y\ mod\ p_1, y\ mod\ p_2, \ldots, y\ mod\ p_n)$ and prove that f is one-to-one and hence onto.

6.4.14 Define the function pp by letting $pp(x)$ be the product of all prime numbers less than x if $x \geq 3$, and by letting $pp(x) = 0$ for $x \leq 2$. Define the function g by $g(x) = pp(ex(2, 2x))$ for all x. (Thus $g(1) = 2 \cdot 3 \cdot 5 \cdot 7 \cdot 11 \cdot 13$. Try writing out $g(2)$!) We shall use two facts about the function g which you can now prove: $ex(3, x) < g(x)$ and $(g(x))^2 < g(x + 1)$ for all $x \geq 1$.

Hint: For very small values of x, these can be proven by direct calculation. For the first of these, you should then use Chebychev's Theorem to prove that $primes(ex(2, x + 1)) > ex(2, x)$ for all $x \geq 3$. For the second inequality, first observe that $2^{x+2} - 4 - x > 2^x + 6 - x$ for all $x \geq 2$, and then use Chebychev's theorem to show that $primes(ex(2, x + 2)) - primes(ex(2, x)) > primes(ex(2, x))$ for all $x \geq 2$.

6.4.15 Prove that for each natural number k there is a formula $N_k(x, y, z)$ in the language of addition which has free variables x, y, and z such that for any natural numbers m, n, and p, the sentence $N_k(\mathbf{m}, \mathbf{n}, \mathbf{p})$ is true if and only if $m \cdot n = p$, $m < g(k)$, and $n < g(k)$. Moreover, there is a positive integer c such that $|N_k| < c(k + 1)$ for all k.

Hints: Suppose that $k>1$, x and y are both less than $g(k)$, and z is less than $g(k + 1)$. Then $x \cdot y$ is also less than $g(k + 1)$, and using Exercise 6.4.13, $x \cdot y = z$ if and only if

$$(x \bmod p \cdot y \bmod p) \bmod p = z \bmod p$$

for all primes p less than $ex(2,2k+2)$. Note that $n < ex(2, 2k)$ if and only if the sentence $M_k(\mathbf{n}, \mathbf{0}, \mathbf{0})$ is true. Give a formula Pr_k such that $Pr_k(\mathbf{p})$ is true if and only if $p < ex(2, 2k)$ and p is a prime. Give a formula G_k such that $G_k(\mathbf{n})$ is true if and only if $n = g(k)$. Give a formula Mod_k such that $Mod_k(\mathbf{x}, \mathbf{y}, \mathbf{z})$ is true if and only if $y < ex(2, 2k)$ and $x \bmod y = z$. Finally, give the formula N_k.

6.4.16 Carefully describe a Turing machine which, for some constant d, on any input $k \geq 0$ produces N_k in time no greater than $d \cdot |N_k| \cdot |k|$. It should run in space $|N_k|$ as well.

6.4.17 Prove that there is an effective translation U which, given any MIN-program $\mathbf{P}$ of n arguments and any $k \geq 0$, produces a formula $U(\mathbf{P}, k)$ in the language of addition with free variables among $x_1, \ldots, x_n$, y such that the sentence $U(\mathbf{P}, k)(\mathbf{m}_1, \ldots, \mathbf{m}_n, \mathbf{q})$ is true if and only if $\mathbf{P}(m_1, \ldots, m_n) = q$ and $\Phi_\mathbf{P}(m_1, \ldots, m_n) < g(k)$. Moreover, for some positive integer d, $|U(\mathbf{P}, k)| < d \cdot |N_k|$. Moreover, show that the translation U is not only effective, but that it is in fact very easy to compute; specifically, show that there is a constant b and a Turing machine which on input $\mathbf{P}$, k produces $U(\mathbf{P}, k)$ in time no greater than $b \cdot |U(\mathbf{P}, k)|^3$.

6.4.18 Define the function t by $t(0) = 0$ and $t(k) = g(k) - 1$ for all $k > 0$. Prove that the theory of addition is rich enough to express the t-limited halting problem for the (complexity measure) Φ defined after Exercise 6.4.3. Conclude that for any Turing machine $\mathbf{d}$ which gives a (partial) decision procedure for the theory of addition there is a constant k and infinitely many true sentences S in the language of addition such that $\text{TMtime}_\mathbf{d}(S) > 2^{2^{|S|/k}}$.

6.4.19 Prove that the constant k in the previous exercise can be fixed, that is, taken to be independent of the Turing machine $\mathbf{d}$.

6.5 DECIDABILITY OF THE THEORY OF ADDITION

In this optional section we give a *brief sketch* of an algorithm for deciding which sentences in the language of addition are true in the usual interpretation. The algorithm has been included for the sake of completeness, both to finish the proof that Φ from the previous section is indeed a complexity measure and also to show that although the theory of addition is intractable, as was shown in the previous section, it is nevertheless algorithmically solvable. We shall assume that the reader is familiar with the fundamentals of mathematical logic in presenting the sketch of the algorithm. The algorithm does provide some illustration of the general theory in that the first step in the algorithm is a "reduction" which (quickly) transforms any sentence in the language of addition into a sentence in another language for limited arithmetic such that the transformed sentence is true (in its usual interpretation) if and only if the original sentence is true. The important concept of reductions is studied at length in the next chapter. The main portion of the algorithm illustrates a common computational procedure in mathematical logic known as "elimination of quantifiers."

We begin by defining $\mathcal{L}'$, a *language for limited arithmetic*, which uses the following symbols:

1. variables: $x, y, z, \ldots$
2. constants: $\mathbf{0, 1}$
3. functions: $\oplus, \ominus$
4. relations: $<, \mid$
5. logical connectives: $\sim, \lor, \&$
6. quantifier: $\exists$
7. punctuation: $(,)$.

$\mathbf{0}$ is a *numeral* of $\mathcal{L}'$, and if $\mathbf{n}$ is a numeral, then so are $(\mathbf{n} \oplus \mathbf{1})$ and $(\mathbf{n} \ominus \mathbf{1})$; if n is *any* integer, positive, negative, or zero, then $\mathbf{n}$ will abbreviate a corresponding numeral. Variables and numerals are *terms* of $\mathcal{L}'$, and if s and t are terms then so are $(s \oplus t)$ and $(s \ominus t)$. If $\mathbf{n}$ is a numeral other than $\mathbf{0}$ and s and t are terms, then $(s < t)$ and $(\mathbf{n} \mid t)$ are (*atomic*) formulas of $\mathcal{L}'$; if x is a variable and F and G are *formulas* of $\mathcal{L}'$, then $\sim F$, $(F \lor G)$, $(F \& G)$, and $\exists x F$ are also formulas. *Free* and *bound* occurrences of variables are defined as usual (see Section 4.1), and the *sentences* of $\mathcal{L}'$ are the formulas with no free variables.

We shall always think of formulas of $\mathcal{L}'$ interpreted in the *usual interpretation* in which variables range over the set of *all* integers $Z = \{\ldots -2, -1, 0, 1, 2 \ldots\}$, $\mathbf{0}$ is 0, $\mathbf{1}$ is 1, $\oplus$ is addition, $\ominus$ is subtraction, $<$ is "less than," and $\mid$ is "divides evenly." Thus "true" and "false" for $\mathcal{L}'$ will always mean true or false, respectively, in this usual interpretation. We shall also use some notational conventions: $m, n, p, \ldots$ will be used to stand for integers, with $\mathbf{m}, \mathbf{n}, \mathbf{p}, \ldots$ standing for the corresponding numerals; Greek letters α, β, . . . will be used to stand for terms; F, G, H, . . . will be used to stand for formulas; parentheses will be omitted whenever possible; and if n is a positive integer and α is a term, $n \cdot \alpha$ will stand for the term $\alpha \oplus \ldots \oplus \alpha$, n times.

6.5.1 EXERCISE Show that the symbol "$|$" is "unnecessary" in $\mathscr{L}'$; that is, given an algorithm which, given any formula of $\mathscr{L}'$, produces another formula of $\mathscr{L}'$ which does not use "$|$" and which is equivalent to the original formula with respect to the usual interpretation. *Hint*: It is important to remember that in an atomic formula of the form $n|\alpha$, n must be a numeral, it is never permitted to be an arbitrary term.

$\mathscr{L}'$ has been carefully chosen to facilitate the "elimination of quantifiers" decision procedure we shall give. The next proposition gives the "reduction" of the language of addition (with its usual interpretation) to $\mathscr{L}'$ (with its usual interpretation).

6.5.2 PROPOSITION *There is an algorithm which given a sentence in the language of addition produces a sentence of $\mathscr{L}'$ which is true if and only if the original sentence is true. Therefore, if there is an algorithm for deciding truth for sentences of $\mathscr{L}'$ then there is one for deciding truth for the language of addition.*

Proof The first part of the transformation is purely "logical": every subformula $\forall x F$ is replaced by $\sim\exists x \sim F$; every subformula $(F \Rightarrow G)$ is replaced by $(\sim F \vee G)$; and every subformula $(F \Leftrightarrow G)$ is replaced by $((F \mathbin{\&} G) \vee (\sim F \mathbin{\&} \sim G))$. In the resulting sentence (which is still in the language of addition), every subformula $\exists x F$ is replaced by $\exists x((x = \mathbf{0} \vee \mathbf{0} < x) \mathbin{\&} F)$ and then every subformula $(\alpha = \beta)$ is replaced by $(\sim(\alpha < \beta) \mathbin{\&} \sim(\beta < \alpha))$. $\square$

Our task now is to show how to decide the truth of sentences in $\mathscr{L}'$. There are *some* sentences in $\mathscr{L}'$ for which it is quite easy to see how to decide whether or not they are true. Specifically, consider those sentences in $\mathscr{L}'$ which contain *no variables*; for example, consider the sentence $(3 \oplus (4 \oplus 10) < 16) \mathbin{\&} \sim(3|8)$. Such sentences simply assert that certain specific integers are or are not less than others and do or do not divide others.

6.5.3 EXERCISE Give the details of an algorithm which, given a sentence in $\mathscr{L}'$ with no variables, decides whether the sentence is true or false.

In the remainder of this section we show how to use the result of the previous exercise to decide whether or not arbitrary sentences in $\mathscr{L}'$ are true in the usual interpretation over $\{\ldots, -2, -1, 0, 1, 2, \ldots\}$. Specifically, given any formula $F(x, \vec{y})$ with free variables $x, \vec{y}$ and with no quantifiers, we show how to produce a formula $G(\vec{y})$ with free variables $\vec{y}$ and with no quantifiers such that for any fixed integer values for the variables $\vec{y}$, $\exists x F(x, \vec{y})$ is true if and only if $G(\vec{y})$ is true.

6.5.4 EXERCISE Show how to use an effective transformation with the properties just described, together with the algorithm from Exercise 6.5.3, to obtain an "elimination of quantifiers" algorithm which decides whether or not arbitrary sentences in $\mathcal{L}'$ are true.

We next give the procedure for eliminating a single quantifier in the following sequence of propositions. If F and G are formulas in $\mathcal{L}'$ with the same free variables, then we say that F and G are *equivalent* if for any fixed integer values for the free variables in them, one is true if and only if the other is true.

6.5.5 PROPOSITION *Let F be a formula in $\mathcal{L}'$ with no quantifiers. Then there is an algorithm which produces an equivalent formula G in $\mathcal{L}'$ with no quantifiers and the same variables as F, and with no negations.*

Proof Using the fact that $\sim(H \vee I)$ is logically equivalent to $(\sim H \& \sim I)$, $\sim(H \& I)$ is logically equivalent to $(\sim H \vee \sim I)$, and $\sim\sim H$ is logically equivalent to H, all negations in F can be "driven in" until all negations occur only immediately in front of atomic formulas. Subformulas of the form $\sim(\alpha < \beta)$ are replaced by $(\beta < (\alpha \oplus 1))$, and subformulas of the form $\sim(\mathbf{n}|\gamma)$ (with $n \neq 0$) are replaced by $(\mathbf{n}|(\gamma \oplus 1) \vee \mathbf{n}|(\gamma \oplus 2) \vee \cdots \vee \mathbf{n}|(\gamma + |\mathbf{n}| - 1))$ where $|n|$ stands for the absolute value of n.

6.5.6 PROPOSITION *Let F be a formula in $\mathcal{L}'$ with no quantifiers, and let x be a variable in F. There is an algorithm which produces an equivalent formula H in $\mathcal{L}'$ with no quantifiers, the same variables as F, and no negations, such that for some positive integer m all occurrences of x in H are in atomic subformulas of one of the following forms: $\mathbf{m} \cdot x < \alpha$, $\beta < \mathbf{m} \cdot x$, or $\mathbf{n}|(\mathbf{m} \cdot x \oplus \gamma)$, where α, β, and γ do not contain x.*

Proof First take G as supplied by Proposition 6.5.5. Then by doing the appropriate additions and subtractions, replace each subformula $(\delta < \sigma)$ which contains x by an equivalent one of the form $(\mathbf{p} \cdot x < \alpha)$ or of the form $(\beta < \mathbf{p} \cdot x)$ for some positive integer p, where α and β do not contain x. Also by doing the appropriate additions and subtractions replace each subformula $(\mathbf{q}|\tau)$ which contains x by an equivalent one of the form $(\mathbf{r}|(\mathbf{s} \cdot x \oplus \gamma))$ for some positive integer s, where γ does not contain x. Next let m (of the statement of the proposition) be a positive *common* *multiple* of all of the integers p and s from the replacements just made. Then some more appropriate additions can be used to replace each atomic formula containing x by an equivalent one in one of the forms required by the proposition.

6.5.7 PROPOSITION *Let F be a formula in $\mathscr{L}'$ with no quantifiers, and let x be a variable in F. There is an algorithm which produces a formula J in $\mathscr{L}'$ with no quantifiers, the same variables as F, and no negations, such that $\exists xF$ and $\exists xJ$ are equivalent and all occurrences of x in J are in atomic subformulas of one of the following forms: $x < \alpha$, $\beta < x$, or $\mathbf{n} \,|\, (x \oplus \gamma)$, where α, β, and γ do not contain x.*

Proof Let H and m be as supplied by Proposition 6.5.6. Let I be obtained by replacing each occurrence of "$\mathbf{m} \cdot x$" in H by "x," and let J be the formula $(\mathbf{m} \,|\, x \,\&\, I)$.

6.5.8 EXERCISE Complete the proofs of the previous three propositions. $\square$

The next proposition finally "eliminates" x.

6.5.9 PROPOSITION *Let F be a formula in $\mathscr{L}'$ with no quantifiers, and let x be a variable in F. There is an algorithm which produces a formula K in $\mathscr{L}'$ with no quantifiers and the same variables as F except for x such that $\exists xF$ and K are equivalent.*

Proof Let x and $\vec{y}$ be the variables in F, and let J be the formula supplied by Proposition 6.5.7. Let J_∞ be the formula obtained by replacing each subformula $x < \alpha$ in J by $1 < 0$ and by replacing each subformula $\beta < x$ in J by $0 < 1$. Then for any fixed integer values for the variables $\vec{y}$, for all sufficiently large values of x, J is true if and only if J_∞ is true. Moreover, if m is now any common multiple of all of the integers n such that subformulas $\mathbf{n} \,|\, (x \oplus \gamma)$ occur in J (which are *all* of the atomic subformulas containing x which occur in J_∞), then for all integers i and j, $J_\infty(\mathbf{i}, \vec{y})$ and $J_\infty(\mathbf{i} + \mathbf{jm}, \vec{y})$ are equivalent. Finally, choosing any such m, we let K be the formula

$$[J_\infty(\mathbf{1}, \vec{y}) \vee J_\infty(\mathbf{2}, \vec{y}) \vee \cdots \vee J_\infty(\mathbf{m}, \vec{y})] \vee$$
$$\bigvee_\alpha [J(\alpha \ominus \mathbf{1}, \vec{y}) \vee J(\alpha \ominus \mathbf{2}, \vec{y}) \vee \cdots \vee J(\alpha \ominus \mathbf{m}, \vec{y})]$$

where the large disjunction $\bigvee_\alpha$ is over all terms α such that $x < \alpha$ is a subformula of J. Clearly K has all of the required properties with the possible exception of being equivalent to $\exists xF$.

Let some integer values for the variables $\vec{y}$ be fixed, and suppose that K is true. If K is true because one of the formulas $J(\alpha \ominus \mathbf{i}, \vec{y})$ is true then $\exists xJ$ is true hence $\exists xF$ is true. If K is true because one of the formulas

$J_\infty(\mathbf{i}, \bar{y})$ is true then $J_\infty(\mathbf{i} \oplus \mathbf{jm}, \bar{y})$ is true for all integers j. But for sufficiently large j, $J_\infty(\mathbf{i} \oplus \mathbf{jm}, \bar{y})$ is true if and only if $J(\mathbf{i} \oplus \mathbf{jm}, \bar{y})$ is true. Thus $\exists xJ$ is true, and so $\exists xF$ is true.

Conversely, let some integer values for the variables $\bar{y}$ be fixed and suppose that $\exists xF$ is true. Then $\exists xJ$ is also true. Let i be an integer such that $J(\mathbf{i}, \bar{y})$ is true. If $J(\mathbf{i} \oplus \mathbf{jm}, \bar{y})$ is true for all positive j then $J_\infty(\mathbf{i} \oplus \mathbf{jm}, \bar{y})$ is true for some sufficiently large j, and hence $J_\infty(\mathbf{i} \oplus \mathbf{jm}, \bar{y})$ is true for all integers j. But then, at least one of the formulas $J_\infty(\mathbf{1}, \bar{y}), \ldots,$ $J_\infty(\mathbf{m}, \bar{y})$ must be true, and hence K is true. On the other hand, suppose that it is not the case that $J(\mathbf{i} \oplus \mathbf{jm}, \bar{y})$ is true for all positive j; let k be such that $J(\mathbf{i} \oplus \mathbf{km}, \bar{y})$ is true but $J(\mathbf{i} \oplus (\mathbf{k} \oplus \mathbf{1})\mathbf{m}, \bar{y})$ is false. Since J has no negations and since the atomic subformulas of J containing x are only in the forms given in Proposition 6.5.7, this can happen only if there is a subformula $x < \alpha$ in J such that $\mathbf{i} \oplus \mathbf{km} < \alpha$ is true and $\mathbf{i} \oplus (\mathbf{k} \oplus \mathbf{1})\mathbf{m} < \alpha$ is false. Therefore, $J(\alpha \ominus \mathbf{p}, \bar{y})$ is true for some p such that $1 \le p \le m$, and hence K is true in this case as well. $\square$

Propositions 6.5.2 and 6.5.9 together with Exercises 6.5.3 and 6.5.4 prove the following theorem.

6.5.10 THEOREM *There is an algorithm for deciding which sentences in the language of addition are true in the usual interpretation.*

The theory of addition is not only a recursive set as shown in the previous theorem. It is elementary. In fact, a careful analysis of the algorithm presented in this section shows that it is 4-elementary, but such an analysis is beyond the scope of this book.

Chapter 7
Complete Problems

Chapter 6 provided examples of computational problems which can be shown to be "intractable," or not "practically solvable," by proving exponential or larger lower bounds on their computational complexity (for infinitely many inputs). Although one might question this interpretation of intractability by asking about average case behavior or about restrictions of the problem to "natural" or "small" inputs, a more detailed mathematical analysis of average case behavior or of the restricted problems has generally resulted in showing that the altered problems are still "intractable" in some very reasonable sense. However, the problems in Chapter 6, and similar problems which have to date been shown to be intractable by the methods of Chapter 6, are not problems which would generally be considered to be vitally important to computing practice.

Other problems, such as sorting, finding the shortest paths between two points in a graph, scheduling tasks, and finding the shortest route connecting a whole collection of points in a graph, occur far more frequently in practice. The current state of knowledge of the complexity of such specific problems has some very large gaps. For some such problems, for example, sorting and finding the shortest paths between two points in a graph, quite good algorithms are known, and these algorithms come reasonably close to attaining the known lower bounds on the complexity of the problems. However, for other problems such as some scheduling problems and finding shortest routes connecting a collection of points in a graph, no nonexponential algorithms are known, and yet no lower bounds beyond the obvious small degree polynomial lower bounds are known.

When the latter situation arises, it is natural and important to attempt to develop better algorithms for the problem, and also to attempt to establish larger lower bounds. If these attempts fail to give satisfactory answers, it may still be possible to use precise mathematical methods to help to explain why the problems are difficult and to provide strong intuitive evidence that the problems are probably intractable. It is the purpose of this chapter to develop such mathematical methods. We begin in Section 7.1 by giving the basic definitions of reducibilities and complete problems illustrated with some examples of the general method. Then in Sections 7.2 and 7.3 we go on to consider important

207

problems from computing practice which the method indicates are probably intractable: the polynomial-space complete problems and the nondeterministic polynomial-time complete problems.

7.1 REDUCIBILITIES AND PROBLEMS COMPLETE IN A CLASS

As a rough example of how one might show that a problem *probably* is difficult to solve computationally without actually *proving* that it is difficult, consider the set RE^2 of regular expressions with squaring representing sets with nonempty complement (see Section 6.2); we shall make this example precise later in this section. Exercise 6.2.10 can rather easily be used to show that *any* set A for which there is an algorithm (e.g., Turing machine) which decides membership in A in space bounded by an exponential function in the length of inputs can be "reduced" to RE^2 in polynomial time, in a sense which implies that if membership in RE^2 *were* decidable in polynomial time then membership in A would also be decidable in polynomial time. Then even if we did not know that RE^2 actually requires exponential space, this alone should convince us that RE^2 "ought" to be very difficult. If we could decide membership in RE^2 in time bounded by a polynomial in the length of inputs, then *every* problem solvable in exponential space would actually be solvable in polynomial time (and hence space); this would be a very unnatural "gap" phenomenon to exist in our natural complexity measures. Since it is surely unlikely that every problem solvable in exponential space can actually be solved in polynomial time (and hence space), it would be reasonable to believe that membership in RE^2 probably cannot be solved in polynomial time or space, even without knowing Theorem 6.2.7. In such a case, we say that every problem in the class $\mathcal{ES}$ of problems solvable in exponential space is "polynomially reducible" to the problem RE^2 in a sense to be made precise later, and this shows that RE^2 is "at least as hard" as any member of $\mathcal{ES}$ to within a polynomial factor. Since by Exercise 6.2.10, RE^2 is in $\mathcal{ES}$, RE^2 is a "hardest," or "complete," member of $\mathcal{ES}$. Thus with the notion of "completeness" made precise, we could have some strong, mathematically exact information about the computational complexity of RE^2 without actually having a lower bound on its complexity: RE^2 is a "hardest" member of $\mathcal{ES}$ and is therefore very probably intractable.

For a somewhat different rough example, take any acceptable programming system and consider the particular recursively enumerable set $K = \{x : \phi_x(x) \text{ is convergent}\}$. As we shall see presently, it is easy to verify that every recursively enumerable (r.e.) set is "algorithmically reducible" to K in a sense which implies that if there were a decision procedure for membership in K then there would also be decision procedures for *all* r.e. sets. Thus, without being certain that K is not recursive, we still might quite reasonably believe that it is unlikely that there is a decision procedure for membership in K. (Since K itself is r.e., we say that K is "complete" in the class of r.e. sets; that is in some sense, K is a "hardest" r.e. set.)

To use the ideas above, we first need to give precise definitions of the notions of reducibility and completeness, and then we can make the above examples precise. Although it is fairly simple to define the notion of one arbitrary function being reducible to another, it is notationally simpler to follow the standard practice of only defining what it means for one set (that is, characteristic function) to be reducible, or to be easily reducible, to another. The most general notion of reducibility for sets is to say that a set A is "algorithmically reducible" to a set B, and to write $A \leq_a B$, if there is some algorithm (or program in some reasonable programming language) which, *if* it were given a subroutine S, which decides membership in B, would *then* decide membership in A; the reduction is called a "polynomial time reduction" if the running time of the algorithm is bounded by a polynomial in the length of the input, *not* counting any time used by the subroutine S. Note that if A is algorithmically reducible to B in polynomial time and if it also happens that the time used by S is bounded by a polynomial in the length of its inputs, then it follows that A can indeed be solved in polynomial time (although this may not seem obvious); similarly, if we know a (very large) lower bound on the complexity of A, we then have a lower bound on the complexity of B which is as large, to within a polynomial factor. This notion of "algorithmic reducibility" is actually broader than we need for this chapter. Consequently, for the purpose of making our ideas mathematically precise we settle on a narrower and simpler definition which is nevertheless adequate for all currently known applications of complexity theory of the type we are considering here.

We say that a set A is *many-one reducible* to a set B, and we write $A \leq_m B$, if there is a total recursive function f such that for all x, $x \in A$ if and only if $f(x) \in B$. If we take the simplistic view that a set is "easy" if it is recursive, and it is "hard" if it is not, then $A \leq_m B$ means that B is at least as hard as A in the following precise sense: *if* S is a subroutine which computes membership in B, *then* we can get an algorithm for membership in A by simply "attaching" S to the end of an algorithm for computing f so that on input x we first compute $f(x)$ and then "consult" the subroutine S to find out whether or not $f(x)$ is in B.

As a good example of many-one reducibility, the reader should consider Proposition 3.2.2. There we proved that for all y and z the set K is many-one reducible to each of the sets $A = \{x : \phi_x$ is a constant function$\}$, $B(y) = \{x : y$ is in the range of $\phi_x\}$, and $C(y, z) = \{x : \phi_x(y) = z\}$.

7.1.1 EXERCISE

(a) Verify the preceding sentence.

(b) Show that $\leq_m$ is a reflexive and transitive relation (that is, that $A \leq_m A$ for all sets A, and if $A \leq_m B$ and $B \leq_m C$ then $A \leq_m C$); do you think that it is symmetric (that is, if $A \leq_m B$ then $B \leq_m A$)?

(c) Show that if $A \leq_m B$ and B is recursive then A is also recursive.

(d) Show that if $A \leq_m B$ and B is r.e. then A is also r.e.

From the preceding exercise we see that if $A \leq_m B$ and membership in A is undecidable, then membership in B cannot be decidable either. Thus with our oversimplified notion of "difficulty," if $A \leq_m B$ then B is "at least as hard" as A. For the reader who is familiar with transfinite cardinalities, it is quite simple to see that there can be no "hardest" set: only countably many of the uncountable totality of all sets (of natural numbers) can be many-one reducible to any particular set. (Why?) Thus, before we can ask the natural question as to whether or not there is a "hardest" set, we must first restrict the class of sets under consideration.

Let $\mathscr{C}$ be any collection of sets. Then a set A is $\leq_m$-*hard for* $\mathscr{C}$ if $B \leq_m A$ for all sets B in $\mathscr{C}$. If A is also in $\mathscr{C}$ then we say that A is $\leq_m$-*complete in* $\mathscr{C}$.

As our first example of a complete set, we shall show that the set K is $\leq_m$-complete in the class of r.e. sets. Historically, this was the first instance of completeness, and the study of complete r.e. sets is an important part of the not-quite-basic theory of the recursive functions.

7.1.2 PROPOSITION   *The set K is $\leq_m$-complete in the class of r.e. sets.*

Proof   Since we already know that K is r.e., we are left with having to prove that K is $\leq_m$-hard for the class of r.e. sets (i.e., $A \leq_m K$ for all r.e. sets A). Let A be any r.e. set and let ψ be a partial recursive function with domain equal to A. By our standard s-m-n construction from Chapter 3, there is a total recursive function f, such that for all x,

$$\phi_{f(x)}(y) = \begin{cases} y & \text{if } x \in A, \\ divergent & \text{if } x \notin A. \end{cases}$$

It is then quite easily verified that for all x, $x \in A$ if and only if $f(x) \in K$.

7.1.3 EXERCISE

(a) Fill in the missing details in the previous proof.   □
(b) Using Exercise 7.1.1 and Proposition 7.1.2, show that $\leq_m$ is not a symmetric relation (i.e., from $A \leq_m B$ it does not necessarily follow that $B \leq_m A$).

Notice that it is only in our oversimplified sense of "difficulty" that it makes sense to interpret Proposition 7.1.2 as showing that K is a "hardest" r.e. set: if K were recursive, then all r.e. sets would be recursive. It does not follow that if membership in K were easily decidable, then membership in any r.e. set would be easily decidable; the function f in the proof of Proposition 7.1.2 could be very hard to compute. However, it does follow that if the function f can be computed in polynomial time and if membership in K could be decided in polynomial time, then membership in any r.e. set could be decided in polynomial time. This, together with examples like Exercise 6.2.10 used in the discussion at the beginning of this section, is some of the motivation behind the following definition.

7.1.4 DEFINITION   We say that a set A is *polynomial time reducible* to a set B, and we write $A \leq_p B$, if there is a total recursive function f such that for all x, $x \in A$ if and only if $f(x) \in B$, and such that the function f can be computed in time bounded by a polynomial in $|x|$ (in any of the standard time measures of Section 1.9). In the literature, this notion of one set being polynomial time reducible to another is often referred to as one set being "polynomially transformable" to another.

7.1.5 EXERCISE

(a) Show that $\leq_p$ is reflexive and transitive, but not symmetric (see Exercise 7.1.1 for definitions).
(b) Show that if $A \leq_p B$ and if there is an algorithm which decides membership in B in time bounded by a polynomial, then there is an algorithm which decides membership in A in time bounded by a polynomial (in any of the standard time measures of Section 1.9).

Throughout this chapter, whenever we discuss computational complexity, it will be with respect to any of the natural time or space measures of Section 1.9 (such as, RAM*time* or TM*space*). From now on, we shall not mention this explicitly but rather assume that it is understood. Of course, even if your favorite time measure is not included in Section 1.9, all of our results will still hold for it as long as it is "natural" enough to be related to one of the time measures in that section by some polynomial. We now make the first example in the discussion at the beginning of this section precise.

7.1.6 PROPOSITION   *Let A be any set for which there is some Turing machine i which decides membership in A and runs in space at most $t_c(|x|)$ on input x, where $t_c(y) = 2^{cy}+y+2$. Also let RE^2 be the set of regular expressions with squaring which represent sets with nonempty complement. Then $A \leq_p RE^2$.*

Proof   We can make a simple modification in i to get a Turing machine j which uses the same portion of tape as i but such that $\phi_j(x)$ is divergent if $\phi_i(x) = 0$ (and $\phi_j(x) = \phi_i(x)$ if $\phi_i(x) = 1$). Then we have that $\mathrm{TM}space_j(x) > t_c(|x|)$ if and only if $x \notin A$, and an application of Exercise 6.2.10 completes the proof.

7.1.7 EXERCISE   Fill in the missing details in the previous proof.   ☐

Thus even if we did not have a proof of an exponential lower bound on the space complexity of RE^2, Proposition 7.1.6 would still be good evidence that membership in RE^2 cannot be decided in polynomial time (or space); RE^2 is to within a polynomial time factor at least as complex as any set whose membership can be decided in exponential space. To finish making this example precise, we need the appropriate notion of completeness.

7.1.8 DEFINITION   Let $\mathscr{C}$ be any collection of sets. We say that a set A is $\leq_p$-*hard for* $\mathscr{C}$ if $B \leq_p A$ for all $B \in \mathscr{C}$. If A is also in $\mathscr{C}$ then we say that A is $\leq_p$-*complete in* $\mathscr{C}$. With the exception of some exercises at the end of this section, for the remainder of this chapter the only reducibility we consider is $\leq_p$ and we shall therefore abbreviate $\leq_p$-hard and $\leq_p$-complete to simply *hard* and *complete*.

If $\mathscr{ES}$ is the class of sets for which membership is decidable in space t_c for some c, then Proposition 7.1.6 shows that RE^2 is hard for $\mathscr{ES}$ and this together with Exercise 6.2.10, shows that RE^2 is in fact complete in $\mathscr{ES}$. This is the precise sense in which RE^2 is (to within a polynomial time factor) a "hardest" set in $\mathscr{ES}$.

As was discussed in the introduction to this chapter, we are interested in studying hard sets for the following reason: there are many computational problems of great interest to computing practice, problems in graph theory, in compiling, in scheduling, etc., for which no nonexponential algorithms are known but for which the only lower bounds on complexity which are known are small degree polynomials. Because of the prevalence of such problems in computing practice, this state of affairs is very unsatisfactory; when such a problem is encountered, one has very little idea as to how much it *should* "cost" to solve it. Thus, in the absence of any reasonably "tight" lower bounds on complexity, if we know that such a problem is complete in some fairly "large" class $\mathscr{C}$

(or is just hard for $\mathscr{C}$) then this is some mathematically precise information about the complexity of the problem which tells us that it is probably "intractable." It turns out that an amazing number and variety of problems are complete in classes $\mathscr{C}$ for which, on intuitive grounds, it seems unlikely to have only members which are solvable in polynomial time.

In Section 7.2 we examine one such class, $\mathscr{PS}$, the class of problems solvable in polynomial space, and in Section 7.3 we will examine a perhaps more restricted class, $\mathscr{NP}$, the class of problems solvable in nondeterministic polynomial time.

It may seem that we have identified polynomial time computability with "tractability," or "practical computability." We are not claiming such an identification. However, there are several good reasons for using it as a reasonable approximation in the general theory of algorithms. First there is the question of using time as the measure of complexity. Nearly all methods for pricing computing services use time as a major basis, and any computation which requires a very large amount of time will result in a very large bill. Thus time measures are our best mathematical tools for dealing with the actual cost of computations. One reason for considering polynomial time bounded algorithms concerns "linear speedup." The history of computer technology is such that every few years there has been a significant drop in computing costs, say by a factor of 10. If an algorithm runs in polynomial time, there is some constant factor (greater than 1) such that every few years the size of problems we can afford to solve with the algorithm is *multiplied* by that constant. For algorithms which require exponential time, such advances result only in a constant amount being *added* to the size of problems we can afford to solve. (For example, $|x|^2=budget$ implies $(\sqrt{10}\,|x|)^2=10 \cdot budget$ while $2^{|x|}=budget$ implies $2^{|x|+\log_2 10} = 10 \cdot budget$.)

There are other reasons for taking the class of polynomials as our time bounds for "practical computability." For some computational problems (such as compiling), a reasonable algorithm should be expected to run in time roughly linear in the length of inputs. For other problems, larger degree polynomial bounds are acceptable. Thus there is no clear cutoff for the degree of polynomial time bounds for "practical computability". Moreover, it is convenient in the general theory of algorithms not to make such a cutoff. The class of polynomials is closed under composition. Therefore, since all reasonable time measures, including those in Section 1.9, are related to each other by polynomial factors, the class of functions computable in polynomial time is the same for all such measures. Although the complexity of *particular* functions in this class could differ significantly from one time measure to another, the *class* of polynomial time computable functions is insensitive to the particulars of a given time measure. Furthermore, we actually use the notion of polynomial time computability to support claims of "intractability" rather than claims of "practical computability". It is generally agreed that a problem which cannot be solved in polynomial time is "intractable." Thus, we are being conservative and safe by taking a class of "practically computable" functions which is very likely too large.

Additional Exercises

*7.1.9 Let Φ be any of the natural time measures from Section 1.9. Let $\mathscr{C}$ be any class of sets, and let A and B be two sets which are both $\leq_p$-complete in $\mathscr{C}$. Suppose that i is an algorithm which decides membership in A (so that the "complexity" of A is at most Φ_i). Show that there is a polynomial p and an algorithm j for deciding membership in B such that for all x there is a y with $\Phi_j(x) \leq \Phi_i(y) + p(|x|)$ and $|y| \leq p(|x|)$.

*7.1.10 Let $\mathscr{C}$ be any class of sets. Show that if $B \in \mathscr{C}$, $A \leq_p B$, and A is $\leq_p$-hard for $\mathscr{C}$, then both A and B are $\leq_p$-complete in $\mathscr{C}$.

*7.1.11

(a) Suppose that $\mathscr{C}$ is a class of sets which is closed under complementation; that is, if A is in $\mathscr{C}$ then $N - A$ (the complement of A) is also in $\mathscr{C}$. Show that if B is $\leq_p$-hard for $\mathscr{C}$ then $N - B$ is also. (*Hint*: First show that for any sets A and B, if $A \leq_p B$ via the function f then $N - A \leq_p N - B$ *via the same function f*.)

(b) Show that both $\mathscr{E}\mathscr{S}$, the class of sets for which membership is decidable in exponential space, and $\mathscr{P}\mathscr{S}$, the class of sets for which membership is decidable in polynomial space, are closed under complementation.

*7.1.12 Prove that the problem of deciding whether two regular expressions with squaring represent the same set is $\leq_p$-complete in $\mathscr{E}\mathscr{S}$, the class of sets for which membership is decidable in exponential space. (*Hint*: Use Exercises 6.2.10, 7.1.10, and 7.1.11, together with Proposition 7.1.6.)

*7.1.13 For all y and z, show that each of the sets $B(y)$ and $C(y, z)$ from Proposition 3.2.2 is $\leq_m$-complete in the class of r.e. sets.

7.1.14

(a) Show that if A is $\leq_m$-complete in the r.e. sets then $N - A$ contains an infinite r.e. subset. Specifically, recall from Exercises 2.4.13 and 3.1.9 that for any acceptable programming system the sets S_0

$= \{x : \phi_x(x) = 0\}$ and $S_1 = \{x : \phi_x(x) = 1\}$ are both r.e., but they are recursively inseparable. Let A be any set which is $\leq_m$-complete in the r.e. sets, and let f be a total recursive function such that $x \in S_0$ iff $f(x) \in A$. Show that $f(S_1) = \{x : x = f(y)$ for some $y \in S_1\}$ is a nonrecursive r.e. subset of $N - A$.

(b) Let Π_1 and Π_2 be projection functions, and define

$$\psi(n) = \Pi_1(min\ z[step(n, \Pi_1(z), \Pi_2(z)) \neq 0 \text{ and } 2n + 1 < \Pi_1(z)])$$

where *step* is a step counting function as in Exercise 3.1.6. (Intuitively, $\psi(n)$ is the first element of the domain of ϕ_n greater than $2n + 1$ which appears when the step counting function is used to enumerate all elements of the domain of ϕ_n.) Let S be the range of ψ; clearly S is r.e. Since $\psi(n) > 2n + 1$ if $\psi(n)$ is defined, $N - S$ is infinite. Show that S is infinite and that $N - S$ has no infinite r.e. subset. (Use the fact that if the domain of ϕ_n is infinite then $\psi(n)$ is defined.) Recursively enumerable sets S for which $N - S$ is infinite with no infinite r.e. subset are called *simple* sets.

(c) Prove that no simple set is $\leq_m$-complete in the class of r.e. sets.

(d) Why are all simple sets nonrecursive?

7.1.15 Define $\leq_1$ just as $\leq_m$ was defined except with the additional requirement that the function f must be one-to-one; define $\leq_1$-*complete in* a class in the obvious way.

(a) Do the appropriate versions of Exercises 7.1.1, 7.1.3, and 7.1.13, with "$\leq_m$" replaced by "$\leq_1$".

(b) Prove that a set is $\leq_m$-complete in the r.e. sets if and only if it is $\leq_1$-complete in the r.e. sets. (*Note:* This part is probably fairly difficult.)

(c) Let S be any simple set, for example, the one defined in Exercise 7.1.14b. Define $S \times N = \{\langle x, y \rangle : x \in S$ and $y \in N\}$. Show that S and $S \times N$ are both r.e. Which of the following hold: $S \leq_m S \times N$, $S \times N \leq_m S$, $S \leq_1 S \times N$, and $S \times N \leq_1 S$? (*Hint:* Draw some pictures.) Prove your answers.

(d) Define $S \times S$ and show that $S \leq_1 S \times S$, but that $S_K \times S_K \leq_m S_K$ is false for S_K defined in Exercise 7.1.16. Nevertheless, give an algorithm which, if given a (fast) subroutine for deciding membership in S, would (quickly) decide membership in $S \times S$.

7.1.16 Let S be the simple set defined in Exercise 7.1.14b, and let $D_n = \{2^n - 1, 2^n, \dots, 2^{n+1} - 2\}$ for all n. Since D_n has 2^n elements and S must have *fewer* than 2^n elements which are less than or equal to $2^{n+1} - 2$, $D_n \cap (N - S)$ is nonempty for all n. Define $S_K = S \cup (\bigcup_{n \in K} D_n)$.

(a) Prove that S_K is simple, and that $K \leq_m S_K$ is false. Show that nevertheless, K is "algorithmically reducible" to S_K since $n \in K$ iff $D_n \subseteq S_K$. Does the reduction you have just given reduce K to S_K in polynomial time? in polynomial space? Prove your answers.

(b) Use either Turing machines or RAM programs together with the intuitive ideas for $\leq_a$ discussed near the beginning of this section to give a careful definition of a general notion "$\leq_a$" of algorithmic reducibility. With your definition, prove that $S \times S \leq_a S$ (see the previous exercise) for all sets S, that $K \leq_a S_K$, that S_K is $\leq_a$-complete in the r.e. sets, and that no recursive set is $\leq_a$-complete in the r.e. sets.

(c) Show that there are nonrecursive r.e. sets which are not $\leq_a$-complete in the r.e. sets. (*Note*: If your definition of $\leq_a$ is correct then this is probably very difficult to prove without some outside help.)

(d) Assuming you have a good definition of $\leq_a$, give a precise definition of "$A \leq_{pa} B$": "A is algorithmically reducible to B in polynomial time." Then verify that if S is a simple set and the pairing function "$\langle \ , \ \rangle$" is reasonable then $S \times S \leq_{pa} S$ while $S_K \times S_K \leq_p S_K$ is false.

(e) Show that there are *recursive* sets A and B, neither of which is empty or N, such that $A \leq_{pa} B$ while $A \leq_p B$ is false.

7.2 POLYNOMIAL SPACE COMPLETE PROBLEMS

In the previous section we proved that RE^2, the set of regular expressions with squaring which represent sets with nonempty complement, is complete in the class $\mathcal{ES}$ of sets for which membership is decidable in exponential space. Although this result does help in better classifying the exact computational complexity of the set RE^2, since in Chapter 6 we had already proven that this problem is exponentially difficult the new knowledge that RE^2 is hard for $\mathcal{ES}$ is not needed to explain why this problem is computationally intractable.

Let $\mathcal{PS}$ be the class of sets for which membership can be decided in space bounded by some polyomial in the length of the input (in any of the standard space measures of Section 1.9). In this section we shall consider some problems which are complete in $\mathcal{PS}$, usually called simply *polynomial space complete* problems. The proofs that the problems of this section are hard for $\mathcal{PS}$, usually called simply *polynomial space hard*, are of interest because all presently known algorithms for these problems require time at least exponential in the lengths of inputs while the largest known lower bounds on their computational time are only small degree polynomials. Thus the knowledge that these problems are polynomial space hard, that is that to within polynomial *time* factors they are at least as difficult as any problem which can be solved in polynomial *space*, is important evidence that very probably there are no algorithms for solving these problems in polynomial time. In fact, to the extent that we believe that not every problem which is solvable in polynomial space is solvable in polynomial time, the knowledge that these problems are polynomial space hard is the best current evidence that these problems are computationally intractable. This knowledge suggests that the gap between our best known algorithms and the largest known lower bounds arises because of our inability to prove exponential time lower bounds for these problems.

Our goal later in this section is to give methods for showing that a number of questions which one might like to answer when compiling high level "procedural" programming languages such as ALGOL-60, questions such as whether various procedures are ever actually used or whether procedures which appear to be recursive (in a sense we shall make precise later) really are recursive, are polynomial space hard even for very simple types of ALGOL-60 programs. However, we first give a simpler example of a polynomial space complete problem building on the work of Chapter 6 and Section 7.1.

Let *RE* be the set of *ordinary* regular expressions (that is, without squaring) which represent sets with nonempty complement. By Exercise 6.2.9, *RE* is in the class $\mathcal{PS}$, and with some appropriate modifications in the proof of Theorem 6.2.7 we shall see that *RE* is also polynomial space hard. Thus we shall have the following:

7.2.1 THEOREM   *The set RE is polynomial space complete.*

Proof   Since we already know that $RE \in \mathcal{PS}$, it remains to show that *RE* is hard for $\mathcal{PS}$. We do this by carefully examining and changing the proof of Theorem 6.2.7, which (very nearly) proves that the corresponding set RE^2 for regular expressions with squaring is hard for $\mathcal{ES}$. In the construction for that proof which precedes Exercise 6.2.6, given a (restricted) Turing machine i and an input $x = x_1 \ldots x_n$ of length n, we constructed a regular expression with squaring $R(i, x)$ such that $\Phi_i(x) > t(|x|) = 2^n + n + 2$ if and only if $L[R(i,x)] = A_k{}^*$. In the construction of $R(i, x)$, squaring was used in only three places: once in "$\{B, \epsilon\}^{2^n}$" and once in "B^{2^n}", both in expression **A"**; and once in "$A_k{}^{2^n}$" in expression **B**. As we saw in Exercise 6.2.3, squaring enables us to write relatively short expressions for sets which would require exponentially longer expressions if squaring were not allowed, and it is just this property of squaring which was exploited in the uses of squaring in the proof of

Theorem 6.2.7. Squaring was used to get expressions of length linear in the length of the input which denote instantaneous descriptions of length exponential in the length of the input. Fortunately, since we only need the reduction to take place in polynomial time, it might well be sufficient to have the expressions be of length polynomial in the length of the input. The proof of Theorem 6.2.7 will now be modified to meet these requirements.

Note that if p is any polynomial then for some fixed m, $p(n) \leq n^m + m + n + 2$ for all n. Thus any Turing machine which runs in polynomial space runs in space bounded by $n^m + m + n + 2$ for some fixed m. For any natural number m, restricted Turing machine i, and input x of length n, define $R(m, i, x)$ to be the *ordinary* regular expression obtained from the expression in the proof of Theorem 6.2.7 by replacing "$\{B, \epsilon\}^{2^n}$" by "$\{B, \epsilon\}^{n^m + m}$" and "$B^{2^n}$" by "$B^{n^m + m}$" in $\mathbf{A}''$, and by replacing "$A_k^{2^n}$" by "$A_k^{n^m + m}$" in $\mathbf{B}$. From the way $R(m, i, x)$ is defined we have that for all x, $\Phi_i(x) > |x|^m + m + |x| + 2$ if and only if $L[R(m,i,x)] = A_k{}^*$. Moreover, from an appropriately modified version of Exercise 6.2.6b we have that for any fixed m and i, $R(m, i, x)$ can be produced (by a Turing machine) in time bounded by a polynomial in the length of x.

Finally, let S be any set for which membership is decidable in polynomial space, and let j be a Turing machine which actually decides membership in S in polynomial space. By making some simple modifications in j we can obtain a restricted Turing machine i such that for some fixed m, if $x \in S$ then $\Phi_i(x) \leq |x|^m + m + |x| + 2$ and if $x \notin S$ then $\Phi_i(x) = \infty$ (that is, i diverges). Thus if we let $f(x) = R(m,i,x)$ then f is computable in polynomial time and $x \in S$ if and only if $f(x) \in RE$. Therefore, $S \leq_p RE$ for all S in $\mathscr{PS}$, and so RE is polynomial space complete.

7.2.2 EXERCISE Give in detail a construction for obtaining machine i from machine j in the last paragraph of the previous proof. $\square$

Our goal for the remainder of this section is to explore the complexity of certain (possibly decidable) questions about the behavior of programs in high level "procedural" programming languages. The programming languages in Chapter 1 are quite low level and, for example, have no explicit facilities for subroutines or "procedures"; such facilities have to be developed by the programmer. On the other hand, the high level languages in which a great deal of practical programming is done usually have rather sophisticated facilities for subroutines or *procedures* which use arguments of various types (integers, reals, arrays, procedures, etc.) called *parameters* to receive inputs and (generally) to return outputs. Appropriate versions of the results we shall give hold for virtually any procedural programming language, but for simplicity we shall deal with a specific, well-known programming language, ALGOL-60. We leave the adaptation of the results to other languages to the reader. We shall not assume that the reader is familiar with ALGOL-60, but we do not want to completely define and develop it either. Instead, we shall give what should be enough examples and discussion for the reader who is not familiar with ALGOL-60 to get a firm grasp of the essential features of the language which are necessary for a full understanding of the results we present.

As a simple example of an ALGOL-60 procedure, consider the following:

```
procedure LS(A, n, L, S);
    integer array A; integer n, L, S;
    begin integer j;
        L := S := A[1];
        for j := 2 step 1 until n do
            begin if A[j] > L then L := A[j];
                  if A[j] < S then S := A[j];
            end;
    end;
```

The procedure LS has parameters A of type "integer array" and n, L, and S of type "integer"; that is, A is expected to be an array of (at least n) integers, and n, L, and S are expected to be integers. A and n serve as inputs to LS, and L and S are used for outputs which, after LS is executed, hold the largest and smallest values, respectively, among the first n entries in A. If after the procedure declaration for LS above an ALGOL-60 program has the instructions

```
LS(B, k, L1, S1);
LS(C, m, L2, S2);
LARGE := L1 − L2;
SMALL := S1 − S2;
```

then the first two instructions are *calls* of the procedure on the arrays B and C, and the last two instructions set the values of the variables LARGE and SMALL to the differences in the largest and the smallest values in the "beginnings" of these two arrays. This illustrates some of the convenience provided by procedures; once LS has been declared as above, simple calls of this procedure can be used many times in a program.

When a program in a high level language is compiled, or translated, into a lower level assembly or machine language prior to execution, for the sake of efficiency it would be desirable to check certain aspects of the program's behavior, provided that the check did not take too long. For example, if a program contains several procedures, it would be very nice to know whether those procedures are actually going to be called, because if not, we need not bother compiling them. Unfortunately, not only can such a check not be made quickly, it cannot be made at all. Consider the following sort of ALGOL-60 program:

```
begin
    integer procedure F(x);
        integer x;
        begin          ⎫
            ⋮          ⎬  "any" ALGOL-60 program
        end;           ⎭
    procedure G(x);
        integer x;
        begin end;
        if F(0) = 0 then G(0);
end.
```

Since the procedure F can compute any partial recursive function ϕ, G will be called in the execution of this program just in case $\phi(0) = 0$, and by Rice's Theorem deciding this is impossible. Therefore, this and almost any other similar behavioral property concerning procedures cannot be tested effectively; it is hopeless to try to determine what will actually happen with procedures by examining a program, say at the time it is compiled.

Of course, it could be that the impossibility of determining what will actually happen with procedures arises because of features of ALGOL-60 other than procedure declarations and procedure calls, for example, arithmetic features such as arithmetic expressions and assignment and conditional statements. Suppose, instead, that we ask a seemingly much weaker question: let us say that a procedure is *formally reachable* if it would be called under the assumption that both possible outcomes of conditional statements take place. Then in the program above, G can clearly be formally reached without knowing anything about procedure F (it is sufficient to see the next-to-last line). It is conceivable that a check for formal reachability could be made at compile time, particularly for "simple" types of programs, and formally unreachable procedures eliminated. However, we shall see that even for ALGOL-60 programs consisting of only procedure declarations and procedure calls the question of formal reachability of procedures is very probably intractable. We shall show this by showing that the problem of formal reachability is polynomial space hard. Of course, for such simple programs (in fact for any programs without conditional statements) the question of formal reachability is the same as the question of "ordinary" reachability. Nevertheless, we have a definite reason for stating the result below in terms of formal reachability. For arbitrary ALGOL-60 programs there is a very simple and natural restriction which is usually satisfied by good programming practice, and which also allows arithmetic expressions and conditional and assignment statements, such that the question of formal reachability of procedures is decidable; in fact, it is decidable in polynomial space. Thus, the formal reachability problem for this restricted class of ALGOL-60 programs is polynomial space complete. For the reader familiar with ALGOL-60 programming, this decidability result is sketched in Exercise 7.2.8.

7.2.3 THEOREM *The problem of whether or not some particular procedure in a given ALGOL-60 program is formally reachable is polynomial space hard, even for programs consisting only of procedure declarations and procedure calls (in particular, with no arithmetic expressions or assignment or conditional statements).*

Proof As in the proof of Theorem 7.2.1, our goal is to show how to reduce a Turing machine which runs in polynomial space to the problem in question. Let S be any set for which membership is decidable in polynomial space. Let p be a polynomial and i be a restricted Turing machine (as in the proofs of Theorems 6.2.7 and 7.2.1) such that if $x \in S$ then $\Phi_i(x) \leq p(|x|)$ and if $x \notin S$ then $\Phi_i(x) = \infty$ but i still does not use more than $p(|x|)$ tape squares. Recall that such a restricted Turing machine never moves more than one square to the left of its starting point, and halts only in a single "halting" state h. Our task now is to write ALGOL-60 programs consisting only of procedure declarations and procedure calls which "simulate" the computation of i on input x.

Suppose that Turing machine i is given input $x = x_1 \ldots x_n$ of length n, and let $m = p(n)$. Since i runs in space at most m and never moves more than one square left of its starting point, tape squares numbered 0 through m are sufficient to keep track of the computation of i on input x. We shall use $m+1$ procedure parameters (which are themselves procedures!) to keep track of the contents of these tape squares. To keep track of i's position on the tape we shall have the tape square under i's read-write head be leftmost, with the other squares arranged as if the tape were made into a circle and then broken to the left of the square under i's read-write head. Thus if square j is being read by i then the tape squares will be in the order $j, j+1, \ldots, m, 0, 1, \ldots, j-1$. Recall that i has states $0, \ldots, h$ with h the "halting" state. To simplify the notation, we shall assume that i uses the two symbol alphabet $\{0(=\text{"blank"}),1\}$; the changes for arbitrary alphabets are obvious. Since we cannot use conditional statements, we must somehow use procedure calls to "test" the tape contents and to perform the appropriate operations on the tape. Suppose, for example, that the instructions for i when in state 0 are 0 0 1 R j and 0 1 0 L k and that the instructions for state $h-1$ are $h-1$ 0 0 L q and $h-1$ 1 1 R r; then the ALGOL-60 program $\mathbf{P}_x$ has procedures STATE 0, $\ldots$, STATE h for i's states, ZERO and ONE for "printing," and LEFT and RIGHT for "moving" as follows:

begin
  procedure STATE0($sq0, \ldots, sqm$);
    procedure $sq0, \ldots, sqm$;
    $sq0(sq0, \ldots, sqm,\text{ONE},\text{RIGHT},\text{STATE}j,\text{ZERO},\text{LEFT},\text{STATE}k)$;
    $\vdots$
  procedure STATE$h-1$($sq0, \ldots, sqm$);
    procedure $sq0, \ldots, sqm$;
    $sq0(sq0, \ldots, sqm,\text{ZERO},\text{LEFT},\text{STATE}q,\text{ONE},\text{RIGHT},\text{STATE}r)$;
  procedure STATEh($sq0, \ldots, sqm$);
    procedure $sq0, \ldots, sqm$;
    begin end;
  procedure ZERO($sq0, \ldots, sqm,pr0,dr0,st0,pr1,dr1,st1$);
    procedure $sq0, \ldots, sqm,pr0,dr0,st0,pr1,dr1,st1$;
    $dr0(pr0,sq1, \ldots, sqm,st0)$;
  procedure ONE($sq0, \ldots, sqm,pr0,dr0,st0,pr1,dr1,st1$);
    procedure $sq0, \ldots, sqm,pr0,dr0,st0,pr1,dr1,st1$;
    $dr1(pr1,sq1, \ldots, sqm,st1)$;
  procedure LEFT($sq0, \ldots, sqm,st$);
    procedure $sq0, \ldots, sqm,st$;
    $st(sqm,sq0, \ldots, sqm-1)$;
  procedure RIGHT($sq0, \ldots, sqm,st$);
    procedure $sq0, \ldots, sqm,st$;
    $st(sq1, \ldots, sqm,sq0)$;
  STATE($x_1, \ldots, x_n,\text{ZERO}, \ldots, \text{ZERO}$);
end.

The program $\mathbf{P}_x$ uses the parameters $sq0, \ldots, sqm$ to "carry" the contents of tape squares, the parameters $pr0$ and $pr1$ to indicate which symbol is to be "printed," the parameters $dr0$ and $dr1$ to indicate the direction to "move," and the parameters $st0$, $st1$, and st to indicate the next state. The instruction in the next-to-last line begins the simulation of i by calling the procedure for state 0 (the initial state) with parameters which specify the initial tape contents. Then a sequence of procedure calls simulates i's computation on input x, ending with a call on procedure STATEh (which does nothing) just in case i halts (that is, just in case x is in S).

In order to get a clearer idea of how this program works, consider the procedures ZERO and ONE. The intuitive interpretation of these procedures is that they expect all of their parameters to be procedures. They each expect the first m of these parameters to be a sequence of the procedures ZERO and ONE which encodes the tape contents of the machine i being simulated. They expect the last three parameters to encode respectively, the tape symbol to be printed, the move to be made, and the state to be entered, if the leftmost procedure (the "symbol" being read) is in fact ONE. They expect the next-to-last three parameters to encode the corresponding information if the leftmost procedure is in fact ZERO. Since they assume that i is positioned over the leftmost parameter, the procedures ZERO and ONE replace the leftmost parameter $sq0$ with which they are supplied by the appropriate procedure representing the symbol to be printed as the first parameter in the procedure call they make. The state to be entered is "stored" in the parameters $st0$ and $st1$, and they pass on the appropriate state as the last parameter in the procedure call they make. Finally, they initiate the appropriate "move" by calling the appropriate move procedure, LEFT or RIGHT.

7.2.4 EXERCISE Let i be the Turing machine in Figure 6.3.4. Pick any reasonable input x and write out the program $\mathbf{P}_x$.

Given i and p, since $m = p(|x|)$ it is trivial to write down program P_x in time bounded by a polynomial in the length of x. Thus, if we define $f(x) = \mathbf{P}_x$ then f is computable in polynomial time and $x \in S$ if and only if in program $f(x)$ the procedure STATEh is formally reachable.   □

The proof of Theorem 7.2.3 can be used to show that many properties of ALGOL-60 procedures in addition to, and perhaps more interesting than, formal reachability are polynomial space hard. In fact, since there is literally nothing to the body of the procedure STATEh in the program $\mathbf{P}_x$, a very wide variety of procedure behaviors and execution errors can be embedded in the program $\mathbf{P}_x$ in the procedure STATEh. Doing so shows that the problems of detecting such behaviors and errors are also polynomial space hard and we shall conclude this section by discussing one such type of procedure behavior, leaving it to the reader to provide additional examples.

One of the useful aspects of ALGOL-60 procedures is the possibility for them to be "recursive"; that is, procedures can be defined in terms of themselves, both directly and indirectly. A simple example of a recursive procedure in ALGOL-60 is the following procedure for computing factorials:

> **integer procedure** FACTORIAL(n);
>   **integer** n;
>   **if** $n \leq 1$ **then** FACTORIAL $: = 1$
>   **else** FACTORIAL $: = n*$FACTORIAL($n-1$);

An instruction containing the procedure call "FACTORIAL(17)" would be supplied with the value 17! by this procedure, which is "directly" recursive in that it contains a call on itself. A procedure can be "indirectly" recursive by containing a call on other procedures which, in turn, can generate calls on the first procedure. For almost any Turing machine i examples of indirectly recursive procedures will result from writing out completely the program $\mathbf{P}_x$ as in Exercise 7.2.4.

Recursive procedures are usually more difficult to implement, or compile, and in general they are more expensive to execute than procedures which are not recursive. Therefore, it would be desirable not to have to implement apparently recursive procedures as if they actually were recursive when they in fact are not. That is, it may seem that a call on a procedure is going to result in the generation of another call on that same procedure before the first call is completed, when in the execution of the program this does not actually take place. If this could be known before the program is compiled, the compiling would be easier and the resulting program would probably be more efficient. Of course, the example discussed prior to the statement of Theorem 7.2.3 can be quite easily adapted to show that the general question of whether a particular procedure is recursive in an ALGOL-60 program is undecidable.

As before, we can consider asking a (seemingly) weaker question: let us say that a procedure is *formally recursive* if a call on it could result in another call on it being generated before the first is completed under the assumption that both possible outcomes of conditional statements take place. Then the factorial procedure above is formally recursive, even though it will actually be recursive only in a program which calls it on a value greater than one. Unfortunately, the question of whether procedures are formally recursive is also polynomial space hard, even for fairly simple types of ALGOL-60 programs. Therefore checking for formal recursiveness of procedures before execution is very probably intractable. Moreover, by appropriately restricting types of ALGOL-60 programs the question of formal recursiveness is decidable, and in fact is polynomial space complete.

7.2.5 COROLLARY   *The problem of whether or not some particular procedure in a given ALGOL-60 program is formally recursive is polynomial space hard, even for programs consisting only of procedure declarations and procedure calls.*

Proof   The proof uses a simple modification of the programs $\mathbf{P}_x$ from the proof of Theorem 7.2.3.

7.2.6 EXERCISE   Prove Corollary 7.2.5 by making appropriate changes in the procedure STATE*h*.   ☐

Theorem 7.2.3 and Corollary 7.2.5 show that testing for reachability or formal recursiveness of ALGOL-60 programs is polynomial space hard, and therefore very probably intractable and almost certainly impractical to do at compile time. Moreover, from the construction of the programs $\mathbf{P}_x$ which are used to prove these results it should be apparent that detecting any property which can be "embedded" in the procedure STATE*h* (or in the procedure STATE*h* in combination with the portion of the program which makes the call on procedure STATE0) without being exhibited in the other procedures of $\mathbf{P}_x$ will also be polynomial space hard. Finally, we point out that although the programs $\mathbf{P}_x$ seem strange because they consist entirely of procedures and have no real input-output behavior, it would be quite easy to attach computations to these programs to make them more reasonable in the sense of actually calculating something. However, by stripping away everything except the procedures themselves, we are left with programs which make clear just how difficult it is to analyze how procedures will actually behave in the execution of ALGOL-60 programs.

Additional Exercises

7.2.7   Change the programs $\mathbf{P}_x$ in the proof of Theorem 7.2.3 so that they remain valid ALGOL-60 programs but so that on execution an incorrect parameter transmission occurs, that is, a procedure is called either on the wrong number of parameters or on parameters of the wrong type, just in case there is a call on the procedure STATE*h*. Conclude that checking for possible parameter transmission errors, say at compile time, is polynomial space hard, even for programs with no arithmetic, assignment, or conditional statements.

7.2.8 In ALGOL-60 it is possible for procedures to use what are sometimes called "global formal parameters"; that is, the body of a procedure can use parameters from other procedures which are not in the formal parameter list and are not declared as variables, parameters, etc., in the procedure in question. It is generally considered that the use of global formal parameters is poor programming practice and should be avoided. In any case, it is quite easy to check whether or not an ALGOL-60 program uses global formal parameters. As an example of the use of global formal parameters, consider and try to analyze the execution of the following ALGOL-60 program which has *x* as a global formal parameter:

```
begin
    procedure A(x);
        procedure x;
        begin
            procedure B(y);
                procedure y;
                y(x);
                x(B);
        end;
    A(A);
end.
```

(a) We need a notion of the "tree of possible executions" for a call on an ALGOL-60 procedure: the root of the tree should represent the procedure call, and for each node of the tree, its immediate descendents should represent other procedure calls which would be generated before the completion of the original call depending on different possible outcomes of conditional statements. Give a precise definition of a tree of possible executions, and use it to give polyomial space bounded algorithms for testing formal reachability and formal recursiveness of procedures in ALGOL-60 programs *without* global formal parameters. Use Theorem 7.2.3 and Corollary 7.2.5 to conclude that both problems are polynomial space complete.

(b) Show that for the class of ALGOL-60 programs without arithmetic, assignment, or conditional statements, but which do allow global formal parameters, the problem of procedure reachability is actually *undecidable*. *Hint*: This part is probably fairly difficult. One way to do it uses nested procedure declarations with formal procedure parameters of outer procedures used inside the bodies of inner procedures as global formal parameters. Repeated calls of the outer procedures can be used to simulate arbitrarily long Turing machine tapes.

7.3 $\mathcal{NP}$-COMPLETE PROBLEMS

In this section we apply the methods of reducibility and completeness to a class of computational problems which includes a large number and wide variety of problems of great importance for computing practice; at present, the completeness of these problems is the best precise mathematical evidence that they are probably intractable, that is, that there probably are no algorithms for solving them in polynomial time.

A basic motivation for studying hard (or complete) problems is that if $\mathcal{C}$ is a presumably large collection of sets which on *general intuitive* grounds based on its definition seems that it should contain sets which are inherently difficult, and if A is a set such that *every* set in $\mathcal{C}$ can easily be reduced to A, then in the absence of any more precise information about the actual computational complexity of A, this is good evidence that A is inherently difficult. In addition, if there are a large number and wide variety of *specific* sets in $\mathcal{C}$ which seem, either by their nature or because they have been studied extensively, to be very difficult, this is further evidence that A is probably inherently difficult. The class of sets $\mathcal{NP}$ which we define later in this section seems to be smaller than $\mathcal{PS}$ yet still possesses each of these features.

In Section 7.1 we saw that the set RE^2 is complete in the class $\mathcal{ES}$ of sets solvable in exponential space; in Section 7.2 we saw that the set RE, along with others, is complete in the smaller class $\mathcal{PS}$ of sets solvable in polynomial space. Since we proved in Chapter 6 that deciding membership in RE^2 actually requires exponential space, no complete problem in $\mathcal{ES}$ can be in $\mathcal{PS}$; thus we know that $\mathcal{PS}$ is properly contained in $\mathcal{ES}$. The exponential difficulty result of Section 6.2 *proves* that any problem hard for $\mathcal{ES}$ requires at least exponential space, and is therefore intractable. In spite of our strong intuition that a polynomial space hard problem ought to require exponential time (the "reusability" of space *ought* to add more computational power), at present there is no known proof that polynomial space hard problems require at least exponential time. For the time being at least, we must live with some uncertainty about the intractability of such problems.

It is certainly plausible that there could be computationally difficult problems, problems requiring exponential time, which are not polynomial space hard. Since we would like to be able to use the method of completeness to obtain information about the computational complexity of as many problems as possible, it is natural to seek a class $\mathcal{C}$ of sets which seems smaller than $\mathscr{PS}$ such that proving that a set is hard for $\mathscr{C}$ would still give us confidence that the problem is probably intractable. With the methods we have developed so far in this book, the only seemingly smaller class which we could easily and naturally define is the class $\mathscr{PT}$ (usually called $\mathscr{P}$ in the literature) of sets for which membership is decidable in time bounded by a polynomial in the length of inputs (in any of the standard time measures of Section 1.9). Certainly, the class $\mathscr{PT}$ is contained in the class $\mathscr{PS}$, but it is certainly not appropriate for our present goal. If a set is complete in $\mathscr{PT}$ then it is solvable in polynomial time, and the fact that every set in $\mathscr{PT}$ is polynomial time reducible to it can hardly be considered evidence that it is intractable; on the contrary, a set which is solvable in polynomial time may well be quite "tractable." In fact, every interesting member of $\mathscr{PT}$ is complete in $\mathscr{PT}$.

7.3.1 EXERCISE Show that every member of $\mathscr{PT}$ with the exception of N and $\emptyset$ is complete in $\mathscr{PT}$. Explain why N and $\emptyset$ are not complete in $\mathscr{PT}$.

Although it is presently unknown whether or not $\mathscr{PT} = \mathscr{PS}$, if $\mathscr{PT}$ does equal $\mathscr{PS}$, then every member of $\mathscr{PS}$ can be solved in polynomial time, which is certainly contrary to the strong intuitions expressed in the previous section and earlier in this section. Therefore it seems worthwhile to look for a class of sets which seems to be intermediate between $\mathscr{PT}$ and $\mathscr{PS}$ and which contains some sets, and hopefully, many sets, which we cannot reasonable believe are in $\mathscr{PT}$. Showing that a set is hard for such a class would show that it is probably intractable even though it may not actually be polynomial space hard. The most important presently known example of such a class is the class $\mathscr{NP}$ of sets solvable "nondeterministically" in polynomial time; it is generally believed at present that this class lies properly between $\mathscr{PT}$ and $\mathscr{PS}$. We shall actually define $\mathscr{NP}$ below, but first we shall discuss the significance of some of the results about problems complete in $\mathscr{NP}$, usually called simply $\mathscr{NP}$-complete problems (similarly, problems hard for $\mathscr{NP}$ are usually called simply $\mathscr{NP}$-hard problems).

There are very many problems of great importance to computing practice which are $\mathscr{NP}$-complete. The $\mathscr{NP}$-complete problems include scheduling problems for computer systems, optimization problems for data bases, a problem called the traveling salesperson problem, and other optimization problems formulated in terms of graphs, integer linear programming problems, and problems of Boolean expression satisfiability. The list of $\mathscr{NP}$-complete problems is very long and constantly growing, and because $\mathscr{NP}$-complete problems all have the same complexity to within polynomial time factors, if any one can be solved in polynomial time, then they all can. Similarly, if any one can be proved not to be solvable in polynomial time, that is not to be in $\mathcal{P}\mathcal{T}$, then none of them can be solved in polynomial time. Thus the question of whether or not $\mathcal{P}\mathcal{T} = \mathcal{N}\mathcal{P}$ is presently one of the most important questions in the theory of computational complexity.

As will be apparent when we define the class $\mathcal{N}\mathcal{P}$, the *general* intuitive basis provided by the definition for believing that $\mathcal{N}\mathcal{P}$ properly contains $\mathcal{P}\mathcal{T}$, while strong, is not so strong as the general intuitive basis for believing that $\mathcal{P}\mathcal{S}$ properly contains $\mathcal{P}\mathcal{T}$. But the large number and variety of $\mathcal{N}\mathcal{P}$-complete problems, many of which have been studied extensively with no success at finding polynomial time algorithms, adds considerable evidence that they are probably intractable.

The class $\mathcal{N}\mathcal{P}$ presently has the following importance for computer science: if a computational problem has consistently resisted attempts to devise "practical" algorithms, then it may well be possible to prove that it is $\mathcal{N}\mathcal{P}$-hard and so probably intractable. Proofs that problems are $\mathcal{N}\mathcal{P}$-hard are often quite easy. Because of the transitivity of the reducibility $\leq_p$, if a set A is known to be hard for some class $\mathcal{C}$ and it can be shown that $A \leq_p B$ for some set B, then it follows that B is also hard for $\mathcal{C}$. Thus to prove that some set B is $\mathcal{N}\mathcal{P}$-hard, it suffices to find any $\mathcal{N}\mathcal{P}$-complete set A among the many which are known and show that $A \leq_p B$. We shall give several applications of this method later in this section.

It is not our purpose in this section to give a lengthy list of $\mathcal{N}\mathcal{P}$-complete problems, or even a reasonably representative sampling. The reader is advised to consult the section on References and Further Readings at the end of the book for books which give a more exhaustive treatment. Our goal here is to give enough examples of $\mathcal{N}\mathcal{P}$-complete problems and of proofs that sets are $\mathcal{N}\mathcal{P}$-hard to illustrate the nature and importance of the class $\mathcal{N}\mathcal{P}$ and the importance of the question of whether or not $\mathcal{P}\mathcal{T} = \mathcal{N}\mathcal{P}$. We also hope to give some idea of the tools which may prove useful for showing that persistently difficult problems are actually $\mathcal{N}\mathcal{P}$-hard.

Before defining the class $\mathcal{N}\mathcal{P}$ we consider a standard and basic example of an $\mathcal{N}\mathcal{P}$-complete problem in order to illustrate some of the ideas involved. A *Boolean expression* consists of variables combined by Boolean, that is logical, connectives such as "not," "and," "or," "implies," etc. We shall consider a restricted type of Boolean expressions; the Boolean expression $F(x_1, x_2, x_3)$ below is in conjunctive normal form:

$$(x_1 \lor \sim x_2) \mathbin{\&} (\sim x_1 \lor x_2 \lor x_3) \mathbin{\&} (\sim x_3).$$

A Boolean expression is in *conjunctive normal form* if it is a conjunction

Table 7.3.2
A Truth Table for the Expression $F(x_1, x_2, x_3)$

Assignment to x_1, x_2, x_3	Resulting Value			
	$(x_1 \vee \sim x_2)$	$(\sim x_1 \vee x_2 \vee x_3)$	$(\sim x_3)$	$F(x_1, x_2, x_3)$
$t \ t \ t$	t	t	f	f
$t \ t \ f$	t	t	t	t
$t \ f \ t$	t	t	f	f
$t \ f \ f$	t	f	t	f
$f \ t \ t$	f	t	f	f
$f \ t \ f$	f	t	t	f
$f \ f \ t$	t	t	f	f
$f \ f \ f$	t	t	t	t

of expressions C_1 & C_2 & . . . & C_n where each *conjunct* C_i is itself a disjunction of expressions $(L_{i_1} \vee \ldots \vee L_{i_m})$, and each *disjunct* L_{i_j} is a *literal*, that is either a variable or the negation of a variable; & stands for "and," $\vee$ stands for "or," and $\sim$ stands for "not." We also make the trivial restriction (which is not usual) that no literal can be used more than once in any particular conjunct. A Boolean expression is *satisfiable* if there is some assignment of the truth values "true" (t) or "false" (f) to its variables so that the whole expression has the value "true." For example, for the formula $F(x_1, x_2, x_3)$ above, $F(t, t, f)$ is "true" and $F(f, f, t)$ is "false."

A standard and straightforward method for testing whether or not a Boolean expression is satisfiable is to build a "truth table" for it. Table 7.3.2 is a truth table for the expression $F(x_1, x_2, x_3)$ above; from it we see that the expression is satisfiable in exactly two ways, by the assignment (t, t, f) and by the assignment (f, f, f). From the computational point of view, building truth tables is unacceptably costly since both the time and space required tend to grow exponentially in the length of the expression. If we are testing for satisfiability, the *space* problem need not be so bad since we simply want to know whether at least one of the assignments yields the value t. Hence, we could test satisfiability in polynomial space by systematically generating the rows one at a time.

7.3.3 EXERCISE

(a) Give a Turing machine which tests satisfiability of Boolean expressions in conjunctive normal form in space bounded by a polynomial in the length of the expressions. In the worst case, what is the running time of the Turing machine in terms of the length of the expression?

(b) Give a Turing machine which tests whether Boolean expressions in conjunctive normal form are *tautologies*, that is, receive value t for all assignments to the variables, in *time* bounded by a polynomial in the length of the expressions.

Since the methods developed so far in this book do not yield a simple and natural way of defining a class of sets which seems to be intermediate between $\mathcal{PT}$ and $\mathcal{PS}$, to define such an intermediate class we shall introduce a different type of model of computation which is definitely and deliberately unrealistic in the sense that it does not correspond to any existing, or imaginable, physical computing device. One of the primary characteristics of computing devices is that they are deterministic: the status of the device at any particular time, including the status of any input channels, completely determines the status of the device at the next moment in time. When actual physical computing devices fail to behave deterministically, they are usually considered to be candidates for repair.

The notion of nondeterministic computation allows there to be more than one choice, or possibility, for the next step at various points in a computation. With nondeterministic computation, one is interested in whether *some* sequence of choices leads to a successful computation (thus there are no probabilities associated with the choices). It turns out that this mathematical fiction is very useful for characterizing a property which many computational problems, including Boolean expression satisfiability, have in common.

The following intuitive comments on nondeterministic computation are intended to be helpful; if they are confusing, they should be disregarded. One way of viewing nondeterministic computation is in terms of "choosing." For example, a Boolean expression is satisfiable just in case there is some choice of truth values for its variables which makes the expression "true," and given any choice of truth values, it is quite easy to compute the value of the expression for that choice. Thus the problem of Boolean expression satisfiability is quite easy to "solve" nondeterministically if some helpful wizard will supply us with a choice of a truth value assignment such that if a particular expression which we are interested in is satisfiable at all then the assignment supplied by the wizard is guaranteed to satisfy the expression; all *we* would have to do is check whether the assignment supplied actually satisfied the expression. Another way this choosing is often explained is in terms of "guessing." If one is a "good guesser" then testing satisfiability is very easy, since after making the best possible guess of a truth value assignment, it is quite easy to evaluate an expression and "know" whether it is satisfiable.

Although it is relatively easy to define nondeterministic versions of almost any programming system, we shall follow the standard practice of using nondeterministic Turing machines. Also, even though it is relatively easy to define the notion of the nondeterministic computation of an arbitrary function, we follow the standard practice of considering only the nondeterministic computation of sets (characteristic functions). Although many computational problems are most naturally represented in terms of functions, it turns out that for almost all computational problems for which nondeterministic computation is relevant there is some closely related set such that solving the computational problem is at least as difficult as deciding membership in the related set. For example, with Boolean expressions it is more natural to ask to be given a satisfying assignment if one exists than simply to be told whether or not one exists. However, solving the first problem algorithmically is certainly at least as difficult as solving the latter, since any solution to the first problem can trivially be converted into a decision procedure for satisfiability with no noticable increase in complexity. Thus, for our purposes, restricting our attention to sets is not a significant limitation.

Recall that when Turing machines were defined at the beginning of Section 1.7 we made them deterministic by requiring that for each state i and symbol a_j there could be at most one instruction beginning "$i\ a_j\ \ldots$". To get a definition of nondeterministic Turing machines, we simply drop this restriction, so that a *nondeterministic Turing machine* may have several instructions which could apply at any point in its computation. A *valid computation* for a nondeterministic Turing machine is simply a sequence of instruction executions such that each instruction executed is one of the instructions which is applicable at that point in the computation. That is, if that machine is in state i reading symbol a_j, then it can choose to execute *any one* of its instructions beginning "$i\ a_j\ \ldots$". With nondeterministic Turing machines we are interested in whether there is *some* valid computation with some particular property. There is a variety of ways of defining what it means for a nondeterministic Turing machine to determine membership in a set, and when restricted to polynomial time they all turn out to be quite obviously equivalent. We shall say that a nondeterministic Turing machine *accepts* an input x just in case some valid computation (beginning in the initial state 0 over the leftmost symbol in x) eventually halts. We say that a nondeterministic Turing machine *accepts* the set S if for all inputs x the machine accepts x if and only if x is in S.

7.3.4 EXERCISE

(a) Outline a method to show that a set is accepted by a nondeterministic Turing machine if and only if it is recursively enumerable.

(b) Suppose that we say that a nondeterministic Turing machine *decides membership in* a set S if every valid computation of the machine eventually halts, and furthermore $x \in S$ if and only if some valid computation on input x halts with a "1" on the tape, and the rest of the tape blank. Show that there is a nondeterministic Turing machine which decides membership in a set if and only if the set is recursive.

Let t be some total (computable) function from N into N. There is a variety of ways to define what it means for a nondeterministic Turing machine to determine membership in a set in time bounded by t, and for "reasonable" functions t they all turn out to be equivalent as well. We say that a nondeterministic Turing machine *accepts* a set S *in time* t if the machine accepts S and if for each $x \in S$ there is some valid computation of the machine on input x which halts after at most $t(|x|)$ instruction executions. A set is *solvable in nondeterministic polynomial time* if there is some nondeterministic Turing machine and some polynomial p such that the machine accepts the set in time p; $\mathcal{NP}$ is the class of sets solvable in nondeterministic polynomial time.

7.3.5 EXERCISE

(a) Let *CNFSAT* be the set of Boolean expressions in conjunctive normal form which are satisfiable. Give a nondeterministic Turing machine which accepts *CNFSAT* in time p for some polynomial p; that is, show that $CNFSAT \in \mathcal{NP}$.

(b) Give an appropriate definition of what it means for a nondeterministic Turing machine to *decide membership* in a set *in time* t (see Exercise 7.3.4b), and show that $\mathcal{NP}$ is the class of sets for which some nondeterministic Turing machine decides membership in polynomial time. (*Hint*: Polynomials are "fairly" easy to compute, so you can build a polynomial "clock" into a nondeterministic Turing machine which terminates any computation when the "clock" runs out.)

There are a number of simple and obvious questions about the class $\mathcal{NP}$ for which the answers are not presently known. It is fairly easy to see that $\mathcal{PT}$ is contained in $\mathcal{NP}$, which in turn is contained in $\mathcal{PS}$, but there are no known proofs that either containment is, or is not, proper. Intuitively, it seems that the "good guessing" aspect of nondeterministic computation would make nondeterministic polynomial time computation more powerful than deterministic polynomial time computation, and that the "reusability" of space would make polynomial space computation more powerful than nondeterministic polynomial time computation, but at this time these only have the status of intuitions. Moreover, while classes such as $\mathscr{PT}$, $\mathscr{PS}$, and $\mathscr{ES}$ are quite clearly closed under complementation, the fact that a set S is in $\mathscr{NP}$ does not seem to necessarily imply that its complement $N-S$ is in $\mathscr{NP}$. It is currently not known whether $\mathscr{NP}$ is closed under complementation, although it is generally doubted.

7.3.6 EXERCISE

(a) Prove that $\mathscr{PT} \subseteq \mathscr{NP}$ and that $\mathscr{NP} \subseteq \mathscr{PS}$. (*Hint*: For the latter use a recursive procedure to test an "accessibility" predicate similar to what was done in Exercise 6.2.9.)

(b) Pick one of the polynomial space complete problems from Section 7.2 and discuss why it does not *seem* to belong to $\mathscr{NP}$.

(c) Try to show that the complement of *CNFSAT* is in $\mathscr{NP}$, and discuss your difficulties. (If you succeed in proving it, please let us know at once.)

Our next goal is to prove that the set *CNFSAT* of satisfiable Boolean expressions in conjunctive normal form is $\mathscr{NP}$-complete. To simplify the proof it is convenient to know that nondeterministic Turing machines are equivalent to restricted ones which can be nondeterministic only in the choice of a next state. A *restricted nondeterministic Turing machine* is a nondeterministic Turing machine with the following properties:

1. The machine has states $0, \ldots, h$ (with 0 the starting state as usual), and h is the only "halting" state; that is, the machine cannot halt in any state other than h and it has no instructions beginning "$h \ldots$", so that when it enters state h it must halt (and thereby accept the input).

2. If $i\, a_j\, a_m\, D\, p$ and $i\, a_j\, a_n\, D'\, q$ are two instructions in the machine, then $a_m = a_n$ and $D = D'$.

7.3.7 PROPOSITION *Any nondeterministic Turing machine can be effectively converted into a restricted nondeterministic Turing machine which accepts the same set and whose nondeterministic running time is at most three times that of the original machine.*

Proof Meeting Condition 1 for a restricted nondeterministic Turing machine is very easy. Meeting Condition 2 can be accomplished by adding two new states for each instruction in the original machine: the first to move the machine back to a square it just left after only being able to choose a new state, the second to perform the appropriate printing and moving before the transition to the next state. (Note that proving this proposition would be easier if we did not require Turing machines to move with each instruction execution.)

7.3.8 EXERCISE Give the details for the proof of the previous proposition. $\square$

7.3.9 THEOREM *The set CNFSAT of satisfiable Boolean expressions in conjunctive normal form is $\mathcal{N}\mathcal{P}$-complete.*

Proof Exercise 7.3.5a showed that $CNFSAT \in \mathcal{N}\mathcal{P}$, so that is remains to show that $CNFSAT$ is $\mathcal{N}\mathcal{P}$-hard. Let S be any set in $\mathcal{N}\mathcal{P}$. By Proposition 7.3.7 there is a restricted nondeterministic Turing machine **T** and a polynomial p such that **T** accepts S in time p. We wish to show that $S \leq_p CNFSAT$. To do this we shall show how, given any x, to write down a Boolean expression B_x in conjunctive normal form which is satisfiable just in case **T** on input x has a valid halting computation of length at most $p(|x|)$; thus we shall have that $x \in S$ if and only if $B_x \in CNFSAT$.

Suppose that **T** has states $0, \ldots, h$ with h the "halting" state, and uses alphabet $A_k = \{a_1, \ldots, a_k\}$ with a_k the blank symbol. For each alphabet symbol a_j and each state $i \neq h$ we let $a_{i,j}$ stand for the unique "symbol to print," a_m; we let $m_{i,j}$ represent the unique "direction to move," D, by letting $m_{i,j} = -1$ if D is L (for left) and letting $m_{i,j} = 1$ if D is R (for right); and we let $ST_{i,j}$ be the set of "possible" next states, that is $ST_{i,j} = \{q:$ there is an instruction $i\, a_j\, a_n\, D\, q$ in the machine$\}$. In addition, since whenever the machine is in state h it just sits there doing nothing, we let $a_{h,j} = a_j$, $m_{h,j} = 0$, and $ST_{h,j} = \{h\}$ for all j.

Let $n = |x|$ and $m = p(n)$. If $x \in S$ then **T** has a valid halting computation of length at most m, and during such a computation **T** can move at most m squares away from its starting point. Thus consecutive tape squares numbered 0 through $2m$ (from left to right) with **T**'s starting point numbered m are sufficient to include all of the tape **T** might need during an accepting computation. Note that the initial contents of this portion of the tape consist of m blanks followed by x followed by $m+1-n$ blanks. Our goal now is to show how Boolean expressions can "simulate" **T**'s computations by writing a Boolean expression B_x in conjunctive normal form which is satisfiable if and only if **T** on input x has an accepting computation of length at most m.

Having fixed x, we use a number of Boolean variables to "describe" the tape contents as well as **T**'s position and state at any point in time up to the limit of $m = p(|x|)$. The variables we use, together with the intended intuitive interpretations of what it means for them to be "true," are as follows:

$SYMB(t, i, j)$ for $0 \le t \le m$, $0 \le i \le 2m$, and $1 \le j \le k$;
                  "true" iff at time t tape position i contains symbol a_j.
$HEAD(t, i)$ for $0 \le t \le m$ and $0 \le i \le 2m$;
                  "true" iff at time t, **T** is at tape position i.
$STATE(t, q)$ for $0 \le t \le m$ and $0 \le q \le h$;
                  "true" iff at time t, **T** is in state q.

Note that the number of these variables is proportional to m^2, and that the maximum length of the "subscripts" in the variables is proportional to the logarithm of m.

Now we shall use the variables above to "describe" **T**'s computations. Since we shall be writing some long conjunctions and disjunctions, we shall use notations such as "$\&_{0 \le i \le m} x_i$" as an abbreviation for "$x_1 \& x_2 \& \ldots \& x_m$." Similarly, if $x_1, \ldots, x_r$ are variables it is convenient to let $JUSTONE(x_1, \ldots, x_r)$ be the Boolean expression

$$[\bigvee_{1 \le i \le r} x_i] \& \&_{1 \le i < j \le r}[\sim x_i \vee \sim x_j].$$

Note that $JUSTONE(x_1, \ldots, x_r)$ is in conjunctive normal form, and that a truth value assignment to $x_1, \ldots, x_r$ makes it true just in case exactly one of the variables is true.

To "simulate" a nondeterministic Turing machine **T**, we first give five Boolean expressions which begin to "assert" that a truth value assignment describes a halting nondeterministic Turing machine computation on the tape squares 0 through $2m$ in time at most m. These expressions actually depend only on the values of m, h, and k, and not on the particulars of **T** and x. The first three expressions say that at each time t, each tape square contains exactly one symbol, the Turing machine is over exactly one square, and the Turing machine is in exactly one state:

(1) $\&_{0 \le t \le m} \&_{0 \le i \le 2m} JUSTONE(SYMB(t, i, 1), \ldots, SYMB(t, i, k))$;

(2) $\&_{0 \le t \le m} JUSTONE(HEAD(t, 0), \ldots, HEAD(t, 2m))$;

(3) $\&_{0 \le t \le m} JUSTONE(STATE(t, 0), \ldots, STATE(t, h))$.

Since we want to keep our expressions in conjunctive normal form, some of them will have to be a bit awkward. Specifically, an expression of the form "$C \& D$ implies E" is equivalent to "$\sim C \vee \sim D \vee E$"; the first is easier to read, while the second is useful for keeping expressions in conjunctive normal form. The following expression says that at each time t if the machine $\mathbf{T}$ is not in position i then the symbol on tape square i cannot change:

(4) $\&_{0 \le t \le m-1} \&_{0 \le i \le 2m} \&_{1 \le j \le k} [HEAD(t, i) \vee {\sim}SYMB(t, i, j) \vee SYMB(t + 1, i, j)]$.

The fifth expression says that at some time t the machine is in state h (thereby halting and accepting the input):

(5) $\bigvee_{0 \le t \le m} STATE(t, h)$.

Any truth value assignment to our variables which satisfies all of these first five expressions must begin to describe a halting, thus accepting, computation of a nondeterministic Turing machine with alphabet A_k and states 0 through h on the tape squares 0 through $2m$ in time at most m. It now remains to give expressions which "assert" that a truth value assignment describes a valid computation of the particular machine $\mathbf{T}$ on input x. The following expression says that the computation starts correctly, that is in state 0 over the left hand end of x with the proper tape contents; let $x = a_{j_1} \ldots a_{j_n}$ with each a_{j_i} a symbol in A_k (recall that symbol a_k is the blank):

(6) $STATE(0, 0) \& HEAD(0, m) \& [\&_{0 \le i \le m-1} SYMB(0, i, k)] \&$
$SYMB(0, m, j_1) \& \ldots \& SYMB(0, m + n - 1, j_n) \&$
$[\&_{m+n \le i \le 2m} SYMB(0, i, k)]$;

The last three expressions say that at each time t, any changes in tape symbols, head position, or state, respectively must be in accordance with $\mathbf{T}$'s instructions.

(7) $\&_{0 \le t \le m-1} \&_{0 \le i \le 2m} \&_{0 \le q \le h} \&_{1 \le j \le k} [{\sim}HEAD(t, i) \vee$
${\sim}STATE(t, q) \vee {\sim}SYMB(t, i, j) \vee SYMB(t + 1, i, a_{q,j})]$;

(8) $\&_{0 \le t \le m-1} \&_{0 \le i \le 2m} \&_{0 \le q \le h} \&_{1 \le j \le k} [{\sim}HEAD(t, i) \vee$
${\sim}STATE(t, q) \vee {\sim}SYMB(t, i, j) \vee HEAD(t + 1, i + m_{q,j})]$;

(9) $\&_{0 \le t \le m-1} \&_{0 \le i \le 2m} \&_{0 \le q \le h} \&_{1 \le j \le k} [{\sim}HEAD(t, i) \vee$
${\sim}STATE(t, q) \vee {\sim}SYMB(t, i, j) \vee \bigvee_{r \in ST_{q,j}} STATE(t + 1, r)]$.

Note that it was precisely Condition 2 for restricted nondeterministic Turing machines (which limits nondeterminism to the choice of next states) which enabled us to use the three separate expressions to describe proper computation steps, and this was very helpful for staying within the bounds of conjunctive normal form.

Now let $B_x = (1) \& (2) \& \ldots \& (9)$. Since each of the expressions (1–9) is in conjunctive normal form, B_x is also in conjunctive normal form.

Moreover, a truth value assignment satisfies B_x if and only if it describes an accepting computation of machine **T** on input x in time at most $p(|x|)$. Thus $x \in S$ if and only if $B_x \in CNFSAT$. To conclude that $S \leq_p CNFSAT$ we need to know that B_x can be produced reasonably quickly, in time bounded by a polynomial in $|x|$. Examination of expressions (1–9) shows that the number of occurrences of variables in each one is either proportional to $m = p(|x|)$ or proportional to $p(|x|)^2$. Thus because of the subscripts in the variables, the length of B_x is at worst proportional to $p(|x|)^3$. Since, given the machine **T** and the input x, the expression B_x is so "simple" it can certainly be written down in time not much greater than its length, say $|B_x|^2$ to be safe. In any case, a Turing machine could surely produce B_x as a function of x in time bounded by a polynomial in the length of x. This completes the proof that $CNFSAT$ is $\mathcal{NP}$-complete.   □

It has taken a lot of work to show that $CNFSAT$ is $\mathcal{NP}$-complete, but it was well worth the effort. As we observed earlier in this section, once we have some sets which we know are $\mathcal{NP}$-complete, it is not necessary to show that every set in $\mathcal{NP}$ is polynomial time reducible to a set A in order to conclude that A is $\mathcal{NP}$-hard; it is enough to show that a single $\mathcal{NP}$-complete set is polynomial time reducible to A.

As our first example of this technique, we establish that a restricted version of the $CNFSAT$ problem is still $\mathcal{NP}$-complete. Recall that a *literal* is either a variable or the negation of a variable.

7.3.10 PROPOSITION   *Let* $3CNFSAT$ *be the set of satisfiable Boolean expressions in conjunctive normal form* $C_1 \& \ldots \& C_n$ *where each conjunct* C_i *is a disjunction of exactly three literals.* $3CNFSAT$ *is* $\mathcal{NP}$-*complete.*

Proof   Given any string x, it is certainly easy to check (deterministically) in time polynomial in the length of x whether x is a Boolean expression in conjunctive normal form with exactly three literals per conjunct. This, together with the fact that $3CNFSAT \subseteq CNFSAT \in \mathcal{NP}$, shows that $3CNFSAT$ is in $\mathcal{NP}$. It remains to show that $3CNFSAT$ is $\mathcal{NP}$-hard. We do this by showing that $CNFSAT \leq_p 3CNFSAT$; this requires a polynomial time computable function f such that $x \in CNFSAT$ if and only if $f(x) \in 3CNFSAT$.

First, we dispose of some trivial cases. For any input x, we begin by testing whether it is a Boolean expression in conjunctive normal form; this is done (deterministically) in time bounded by a polynomial in $|x|$. If x does not pass the test then we let $f(x) = x$. If x does pass the test then we need to produce "quickly" a Boolean expression $f(x)$ with exactly three literals per conjunct which is satisfiable if and only if x is satisfiable.

Let x be $C_1 \& \ldots \& C_n$. It will suffice to show how to "quickly" convert each C_i into an "equivalent" Boolean expression in conjunctive normal form with exactly three literals in each conjunct. If C_i happens to be a disjunction of fewer than three literals, then it is quite easy to expand it into an "equivalent" expression of the required type. For example, if C_i is $(L_1 \lor L_2)$ and y is a new variable (not in x), then C_i can obviously be replaced by the expression $(L_1 \lor L_2 \lor y) \& (L_1 \lor L_2 \lor \sim y)$. A conjunct of the form (L_1) can be replaced by

$$(L_1 \lor y \lor z) \& (L_1 \lor \sim y \lor z) \& (L_1 \lor y \lor \sim z) \& (L_1 \lor \sim y \lor \sim z).$$

For the more general case when C_i is $(L_1 \lor \ldots \lor L_k)$ with $k \geq 4$, let $y_1, \ldots, y_{k-3}$ be new variables. Then C_i can be replaced by the following expression:

$$(L_1 \lor L_2 \lor y_1) \& (L_3 \lor \sim y_1 \lor y_2) \& (L_4 \lor \sim y_2 \lor y_3) \& \ldots$$
$$\& (L_{k-2} \lor \sim y_{k-4} \lor y_{k-3}) \& (L_{k-1} \lor L_k \lor \sim y_{k-3}).$$

Any assignment which makes C_i true must make some L_j true, and then making y_m true for $m < j - 1$ and y_m false for $m \geq j - 1$ makes the expression above come out true. Conversely, suppose for the sake of a contradiction that some assignment made the expression above true but made C_i false. Then each L_j must be false under this assignment, and by examining the first conjunct we see that the only way for the expression above to be true is if y_1 is true. Proceeding to the second conjunct, we see that then y_2 must also be true. By the time we reach the next-to-last conjunct we see that all of the y's must be true. This leaves us with the last conjunct being false, contradicting our assumption that the expression is true under the given assignment. Therefore any assignment which makes the expression above true must also make C_i true.   □

With Boolean expressions, we are concerned with satisfying assignments. Our next $\mathcal{NP}$-complete problem, as is the case with many others, comes from an optimization problem. In the *knapsack optimization problem*, we are given a sequence of $n+1$ integers consisting of *sizes* $s_1, \ldots, s_n$ for n *items* followed by a *capacity* c for a *knapsack*; we wish to find a subset of the set of items which comes as close as possible to filling the knapsack without overflowing, that is, so that the sum of the sizes in the subset is no more than c. For one possible application, let the items be goals, the sizes be the costs of attaining the goals, and c be the total budget available. In the *knapsack question*, we are also given a set of items with corresponding sizes and a knapsack with some capacity, but we simply wish to know whether some subset of the items exactly fills the knapsack. The knapsack question is at least as hard to solve as the knapsack optimization problem in the following precise sense: any algorithm which solves the knapsack optimization problem can easily be converted into one which answers the knapsack question with no significant increase in complexity. Therefore, if the knapsack question is $\mathcal{NP}$-complete then there is probably very little hope for efficient solutions for the knapsack optimization problem.

7.3.11 THEOREM   *Let KNAPSACK be the set of knapsack questions for which the answer is "yes;" that is, the sum of some subset of the sizes is the capacity. KNAPSACK is $\mathcal{NP}$-complete.*

Proof   It is easy to see that $KNAPSACK \in \mathcal{NP}$. Somewhat informally, given a knapsack question, simply pick (nondeterministically) a subset of the sizes, add them up, and see whether the total equals c. To show that $KNAPSACK$ is $\mathcal{NP}$-hard, we show that $3CNFSAT \leq_p KNAPSACK$. Since this is our first example of reducing one question to a rather different sort of question, we discuss it in fairly complete detail.

To show that $3CNFSAT \leq_p KNAPSACK$ we need to produce a function f which is computable in polynomial time such that $x \in 3CNFSAT$ if and only if $f(x) \in KNAPSACK$. To dispose of trivial cases, we first check whether x is a Boolean expression in conjunctive normal form with three literals per conjunct. If x is not such an expression, then we let $f(x)$ represent the knapsack question with $s_1 = 1$ and $c = 0$. If x is such an expression, then we want $f(x)$ to represent a knapsack question with answer "yes" just in case x is satisfiable. We shall actually produce a knapsack question such that any satisfying assignment for x corresponds directly to a set of items which exactly fill the knapsack, and *vice versa*.

Suppose that x is $C_1 \& \ldots \& C_n$ and contains the variables $v_1, \ldots, v_m$. The simplest idea is somehow to give sizes $s_1, \ldots, s_m$ for the variables and pick a capacity c so that the variables made true in a satisfying assignment correspond to a set of sizes whose sum is c. But this idea does not seem to work, because although a true variable could contribute to filling the knapsack in accordance with the conjuncts which it makes true, a false variable may also make some conjuncts true, and under such a scheme it would be unable to contribute to filling the knapsack. Therefore it seems necessary to use the next simplest idea; we give sizes for all literals, that is one size for each variable and one size for the negation of each variable. We would like to do this and to pick a capacity c so that the literals made true in a satisfying assignment correspond to a set of sizes whose sum is c.

In order to have sets of sizes which fill the knapsack correspond to satisfying assignments it is necessary to insure that in any such set of sizes, for each variable v_i the set contains exactly one of the sizes corresponding to v_i or to $\sim v_i$. One way to accomplish this is to use the rightmost m digits in the sizes as follows: in each of the sizes corresponding to v_i and to $\sim v_i$ the ith digit (from the right) is a 1 and the remaining $m-1$ digits are 0. Then if we let the rightmost m digits of the capacity c be 1, any subset of the sizes whose sum equals c must have m members such that for each variable v_i the set contains one size corresponding either to v_i or to $\sim v_i$, but not both; conversely, any such set of m sizes will have its rightmost m digits all 1's, that is equal to the rightmost m digits in c.

The rightmost m digits in the sizes and in c insure that sets of sizes whose sum equals c correspond to partitions of the literals in the expression x into two subsets such that for each variable v_i, v_i and its negation are each in separate parts of the partition. We now add n more digits to the sizes, one corresponding to each of the conjuncts $C_1, \ldots, C_n$. For literal L_i the $m+j$th digit (from the right) in its corresponding size is a 1 if L_i is in C_j, otherwise it is a 0. The top portion of Table 7.3.12 illustrates the sizes corresponding to the literals for the example when C_j is $(v_1 \bigvee \sim v_i \bigvee v_m)$ and C_k is $(v_1 \bigvee \sim v_1 \bigvee v_i)$. Since we have defined the digits in the size corresponding to each literal L_i so that the $m+j$th digit is a 1 just in case L_i being true makes conjunct C_j true, satisfying assignments to the variables correspond to sets of sizes whose sum is an $n+m$ digit integer with its rightmost m digits all 1's and its leftmost n digits *at least* 1 and *at most* 3. But since we need to fill the knapsack exactly, this does not quite work.

When we recall that each conjunct contains exactly three literals, it is quite easy to see how to arrange to fill the knapsack exactly. We add $2n$ more "compensating" sizes, two corresponding to each conjunct C_j with the $m+j$th digit a 1 and the remaining digits 0's. Table 7.3.12 illustrates these compensating sizes along with those corresponding to the literals. Now suppose we have a satisfying assignment for the expression x and we take the sizes corresponding to the literals made true by the assignment, and in addition, for each conjunct C_j with only one or two of its literals made true by the assignment we take two or one (respectively) of the compensating sizes corresponding to C_j. Then the sum of this set of sizes is the $m + n$ digit number 33 . . . 3311 . . . 11 consisting of n 3's followed by m 1's, and this is the value we take for the capacity c of the knapsack. Thus, for any satisfying assignment for x, there is a corresponding set of sizes whose sum is c.

Conversely, suppose that some set of sizes has its sum equal to c. The last m digits of c insure that this set contains one size corresponding to

Table 7.3.12

Literal and Compensating Sizes for the Expression $x = C_1 \ \& \ \ldots \& \ C_n$ **Using Variables** $v_1, \ldots, v_m$; $C_j = (v_1 \vee \sim v_i \vee v_m)$ **and** $C_k = (v_1 \vee \sim v_1 \vee v_i)$

	DIGITS							
	$m+n$	..	$m+k$	..	$m+j$	..	$m+1$	$m \ldots i \ldots 2\ 1$
v_1	?	..	1	..	1	..	?	$0 \ldots 0 \ldots 0\ 1$
$\sim v_1$	?	..	1	..	0	..	?	$0 \ldots 0 \ldots 0\ 1$
v_2	?	..	0	..	0	..	?	$0 \ldots 0 \ldots 1\ 0$
$\sim v_2$	?	..	0	..	0	..	?	$0 \ldots 0 \ldots 1\ 0$
$\vdots$			$\vdots$					
v_i	?	..	1	..	0	..	?	$0 \ldots 1 \ldots 0\ 0$
$\sim v_i$	?	..	0	..	1	..	?	$0 \ldots 1 \ldots 0\ 0$
$\vdots$			$\vdots$					
v_m	?	..	0	..	1	..	?	$1 \ldots 0 \ldots 0\ 0$
$\sim v_m$	?	..	0	..	0	..	?	$1 \ldots 0 \ldots 0\ 0$
C_1	0	..	0	..	0	..	1	$0 \ldots 0 \ldots 0\ 0$
C_1	0	..	0	..	0	..	1	$0 \ldots 0 \ldots 0\ 0$
$\vdots$			$\vdots$					$\vdots$
C_j	0	..	0	..	1	..	0	$0 \ldots 0 \ldots 0\ 0$
C_j	0	..	0	..	1	..	0	$0 \ldots 0 \ldots 0\ 0$
$\vdots$								$\vdots$
C_k	0	..	1	..	0	..	0	$0 \ldots 0 \ldots 0\ 0$
C_k	0	..	1	..	0	..	0	$0 \ldots 0 \ldots 0\ 0$
$\vdots$			$\vdots$					$\vdots$
C_n	1	..	0	..	0	..	0	$0 \ldots 0 \ldots 0\ 0$
C_n	1	..	0	..	0	..	0	$0 \ldots 0 \ldots 0\ 0$

each variable or its negation, but never both; take the truth value assignment which makes each literal corresponding to one of the sizes in the set true. For each conjunct C_j, the sum of the $m+j$th digits in the sizes in the set equals 3 and at most 2 could come from compensating sizes. Therefore there is at least one size corresponding to a literal in the set with a 1 as its $m+j$th digit. Since this corresponding literal is true, the conjunct C_j is also true, and thus we have a satisfying assignment for the expression x.

Since the knapsack question above can clearly be produced in time polynomial in the length of x, this completes the proof that $3CNFSAT \leq_p KNAPSACK$. $\square$

A problem which is very closely related to the knapsack question is the *partition question*, in which we are given a sequence of n integers $s_1, \ldots, s_n$ and we wish to know whether there is some subset of the set of integers whose sum is exactly half of the sum of all of the integers. Obviously, the partition question is a restriction of the knapsack question in which the capacity c is required to be half of the sum of all the sizes $s_1, \ldots, s_n$. A fairly simple modification of the proof that *KNAPSACK* is $\mathcal{NP}$-complete shows that this restriction is also $\mathcal{NP}$-complete.

7.3.13 COROLLARY *Let PARTITION be the set of partition questions for which the answer is "yes." PARTITION is $\mathcal{NP}$-complete.*

7.3.14 EXERCISE Prove the previous corollary. If you need a hint, see the restatement of this exercise at the end of this section.

Another optimization problem similar to the knapsack problem is the *general scheduling problem*: we are given k *processors* and a sequence of n *jobs* requiring *times* $t_1, \ldots, t_n$ to be completed (the times can be real numbers and there may be some restrictions placed on the order in which some of the jobs can be done), and we want to schedule the jobs for the k processors so that we finish processing all of the n jobs as quickly as possible. In the *scheduling question* we are given k processors, n jobs, integer times $t_1, \ldots, t_n$, and an integer *time bound b*, and we wish to know whether the jobs can be split into k groups so that the sum of the times for the jobs in each group is at most b. The general scheduling problem is clearly at least as difficult as the scheduling question. Also notice that the partition question is simply a restriction of the scheduling question with two processors and a time bound which is as tight as possible without obviously ruling out a solution.

7.3.15 COROLLARY *Let SCHEDULING be the set of scheduling questions for which the answer to the scheduling question is "yes." SCHEDULING is $\mathcal{NP}$-complete.*

7.3.16 EXERCISE Using Corollary 7.3.13, prove the previous corollary; if it takes more than two minutes, you have used too much time.

The next problem we consider concerns what are known as Hamilton paths in directed graphs. A *directed graph* consists of some points, or *vertices*, and some arrows, or *edges*, connecting pairs of distinct vertices. A *path* in a directed graph is a sequence of edges such that the "head" of each edge (except the last) connects to the "tail" of the next edge in the sequence, and such that no vertex is "visited" twice by the path. A *Hamilton path* is a path which visits all vertices in the directed graph. Figure 7.3.17 shows two directed graphs; the first has two Hamilton paths while the second has none. In the *Hamilton path problem* we are given a directed graph G and two vertices v and w in G, and we wish to find a Hamilton path in G from v to w if one exists. In

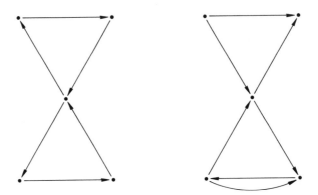

FIGURE 7.3.17 Two directed graphs, one with Hamilton paths, one without.

the *Hamilton path question* we simply wish to know whether or not there is a Hamilton path from v to w. Obviously the Hamilton path problem is at least as hard as the Hamilton path question.

7.3.18 THEOREM   *The Hamilton path question is $\mathcal{NP}$-complete.*

Proof First notice that we have not specified a method for using strings, that is integers to some base $k \geq 2$, to represent directed graphs. In fact, for our purposes here any reasonable method will do; it is only necessary that the length of a string representing a directed graph be bounded by a polynomial in the total number of vertices and edges in the graph, and that we be able to reconstruct the graph from its representation in time polynomial in the length of the representation. It would require some careful thought to produce a representation which does not meet these conditions.

Showing that the Hamilton path question is in $\mathcal{NP}$ is just as simple as showing the other $\mathcal{NP}$-complete questions we have seen are in $\mathcal{NP}$. Given a directed graph G with n vertices and vertices v and w in G, we simply choose (nondeterministically) a set of $n-1$ edges in G and check (deterministically) whether these edges form a path from v to w. If so, there is a Hamilton path from v to w in G, since any path with $n-1$ edges must visit n distinct vertices.

We show that the Hamilton path question is $\mathcal{NP}$-hard by showing that 3CNFSAT is polynomial time reducible to it. Given x, if x is not a Boolean expression in conjunctive normal form with three literals per conjunct then we let $f(x)$ be (that is, represent) any directed graph without Hamilton paths, say the second one in Figure 7.3.17. If x is such a Boolean expression, we shall show how to "quickly" construct a directed graph G with a pair of vertices v and w such that there is a Hamilton path from v to w in G just in case x is satisfiable.

To see how to construct the graph G we shall for the present *assume* that we have a section S of a directed graph with some special properties. A section S of a directed graph is called a *three lane carriageway* (or *TLC* for short) if S has three input edges i_1, i_2, i_3 and three corresponding output edges o_1, o_2, o_3 such that S has the following properties:

1. Given any set of one, two, or three input edges of S there are one, two, or three paths (respectively) in S beginning with these edges and which between them visit all vertices in S exactly once, and such that each path ends with the output edge o_j corresponding to the input edge i_j with which it started.
2. For any set of one, two, or three input edges of S there is only one set of one, two, or three paths (respectively) in S beginning with these edges which between them visit all vertices in S exactly once and with each path ending in *some* output edge of S.

Figure 7.3.19a shows the form of a *TLC*, and Figure 7.3.19b–d shows some possible sets of paths through it illustrating properties 1 and 2. In Figure 7.3.22 we shall give an actual *TLC*.

Let $x = C_1 \& \ldots \& C_m$ be a Boolean expression in conjunctive normal form with three literals per conjunct which uses variables $y_1, \ldots, y_n$. We shall build a directed graph G out of m *TLC*'s (corresponding to x's m conjuncts) and $n+1$ additional vertices $v_1, \ldots, v_{n+1}$ such that there

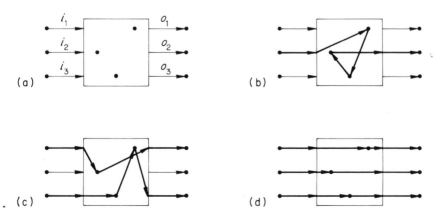

FIGURE 7.3.19 The form of a *TLC* and some possible sets of paths through it.

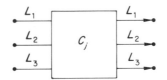

FIGURE 7.3.20 A *TLC* corresponding to a conjunct C_j of the form $L_1 \vee L_2 \vee L_3$).

is a correspondence between satisfying assignments for x and Hamilton paths from v_1 to v_{n+1} in G. Suppose that conjunct C_j in x is $(L_1 \vee L_2 \vee L_3)$. For each such conjunct, its corresponding *TLC* will have its input and output edges identified with L_1, L_2, L_3 as in Figure 7.3.20.

For any variable y_i in x let $B_{i,1}, \ldots, B_{i,j}$ be the conjuncts of x which contain the literal y_i and let $D_{i,1}, \ldots, D_{i,k}$ be the conjuncts which contain the literal $\sim y_i$. Then in the directed graph G the *TLC*'s corresponding to these conjuncts and the vertices v_i and v_{i+1} are connected as in Figure 7.3.21. When this is done for all the variables $y_1, \ldots, y_n$, then each *TLC* will appear in three such pictures and there will be no "dangling edges" left on any of the *TLC*s.

Suppose that the directed graph G has a Hamilton path from v_1 to v_{n+1}. From the way G is constructed it follows that the Hamilton path includes exactly one of the edges leading out of each vertex v_i for $1 \le i$

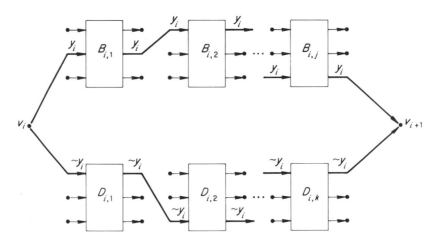

FIGURE 7.3.21 Connections in G corresponding to the literals y_i and $\sim y_i$.

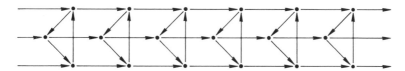

FIGURE 7.3.22   A *TLC*.

$\leq n$, and from the properties we have assumed for *TLC*'s it follows that the path includes either all of the edges identified with the literal y_i and none identified with $\sim y_i$ or it includes all of the edges identified with $\sim y_i$ and none identified with y_i. Take as our truth assignment for the variables $y_1, \ldots, y_n$ that assignment which makes those literals identified with edges in the Hamilton path true. Since the Hamilton path must go through each *TLC* and since the truth assignment makes each literal identified with an edge on the path true, the truth assignment makes at least one literal in each conjunct true. Therefore the truth assignment makes each conjunct true, and so it makes x true.

Conversely, suppose that x has a satisfying assignment. We start to build a Hamilton path by choosing all edges in G which are identified with literals made true by the assignment. Thus we have chosen exactly one of the edges leading out of each vertex v_i for $1 \leq i \leq n$ and exactly one of the edges leading into each vertex v_{i+1} for $1 \leq i \leq n$, and for each conjunct C_j in x we have chosen some set of one, two, or three input edges and the corresponding output edges in the *TLC* which corresponds to that conjunct. Then by the properties we have assumed for *TLC*'s, there is exactly one way to complete this path and get a Hamilton path from v_1 to v_{n+1} in G.

Therefore, assuming that there really is such a thing as a *TLC* we have shown that $3CNFSAT$ is polynomial time reducible to the Hamilton path question. (Note that this is true no matter how large and complicated a *TLC* might be, since we would be copying the same one repeatedly in constructing the graph G.)

To finish the proof that the Hamilton path question is $\mathcal{NP}$-complete, we reach into our large black hat and pull out the *TLC* in Figure 7.3.22.

7.3.23 EXERCISE   Verify that Figure 7.3.22 is in fact a *TLC*. To check the case when you have a set of two input edges, you may find it helpful to scribble on a copy of the *TLC*, but not on the one in this book. $\square$

An optimization problem which has been studied at great length and which is closely related to the Hamilton path problem is called the traveling salesperson problem. In a directed graph, a *walk* is like a path except that it is allowed to reuse vertices and edges. Suppose that we are given a directed graph G containing vertices v and w, and suppose that there are *lengths* (integer or real) associated with the edges of G. In the *traveling salesperson problem* we wish to find the shortest possible walk from v to w which visits all the vertices in some designated subset of G. In the *traveling salesperson question* the lengths are integers and we are also given an integer *bound* b, and we wish to know whether there is some walk from v to w visiting every vertex in some designated subset of G and having length at most b. Obviously the traveling salesperson problem is at least as hard as the traveling salesperson question. In addition, the Hamilton path question is just the restriction of the traveling salesperson question to the case in which all the edges have length 1, the bound is one less than the number of vertices in the graph, and the designated subset consists of all vertices in G.

7.3.24 COROLLARY   *The traveling salesperson question is $\mathcal{NP}$-complete.*

7.3.25 EXERCISE   Prove the previous corollary.   □

In this section we have seen several $\mathcal{NP}$-complete questions, most of them arising from optimization problems of great importance to computing practice; a "fast" algorithm for one of these optimization problems would provide anxiously awaited feasible computer solutions to currently "intractable" problems. Unfortunately, since all $\mathcal{NP}$-complete questions have the same difficulty to within polynomial time factors, and since there are a wide variety and large number of $\mathcal{NP}$-complete problems (with the list presently running into the thousands), it seems very unlikely that there are "fast" algorithms for any $\mathcal{NP}$-complete problems.

Since there are many $\mathcal{NP}$-complete problems of great importance for computing practice, there are many people faced with a real need to solve such problems. This raises the question of what computational alternatives to "fast" algorithms are available to attempt the feasible solution of $\mathcal{NP}$-complete problems. One possible answer is to look for fast algorithms which yield good approximate solutions to optimization problems. For some problems, scheduling for example, this approach has been successful. Another approach is to restrict the class of inputs in some way that makes the problem easier without ruling out any of the particular problems we actually wish to solve. At present, this approach has not yielded very encouraging results. A third approach is to look for "heuristic" algorithms which produce reasonably good solutions fairly quickly most of the time. Some very recent work has combined these last two approaches; it has yielded a fast algorithm which very probably finds Hamilton paths when ones exist in a restricted class of graphs which are very likely to have Hamilton paths. Thus even if $\mathcal{NP} \neq \mathcal{PT}$ there may still be some reasonable sense in which some $\mathcal{NP}$-hard problems are "practically computable" although not polynomial time computable.

Additional Exercises

*7.3.14

(a) Prove that *PARTITION* is $\mathcal{NP}$-complete. (*Hint*: In the reducibility construction in the proof of Theorem 7.3.11 (showing $3CNFSAT \leq_p KNAPSACK$), the sum of all sizes used is $55 \ldots 5522 \ldots 22$; that is, n 5's followed m 2's. By adding $11 \ldots 1100 \ldots 00$ (n 1's followed by m 0's), the sum becomes $66 \ldots 6622 \ldots 22$. Then satisfying assignments correspond to the halves of successful partitions which do not contain the newly added size.)

(b) By showing that the partition question is just a special case of the knapsack question, give a very easy proof that *PARTITION* $\leq_p$ *KNAPSACK*. Conclude that Theorem 7.3.11 is an easy corollary of Exercise 7.3.14a.

7.3.26 An (*undirected*) *graph* is like a directed graph except that the edges are plain arcs instead of arrows (they have no "direction").

(a) A *path* in an undirected graph is just like one in a directed graph except, of course, that the heads and tails of arrows do not have to match up; a *Hamilton path* is a path which visits every vertex. Show that the Hamilton path question for undirected graphs is

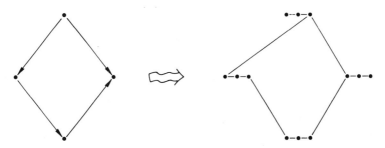

FIGURE 7.3.27   A big hint.

$\mathcal{NP}$-complete by giving a polynomial time reduction of the Hamilton path question for directed graphs to it. (*Hint*: See Figure 7.3.27.)

(b) A *k-clique* in an undirected graph is a set of k vertices such that the graph has an edge between each pair of distinct vertices in the set. Show that the question of whether a graph G has a k-clique, for arbitrary G and k, is $\mathcal{NP}$-complete. (*Hint*: $3CNFSAT$ is polynomial time reducible to this clique question; try constructing an undirected graph with one vertex for each occurence of a literal.)

7.3.28 Most reductions of one $\mathcal{NP}$-complete problem to another have the property that they preserve the number of solutions to the problem in question. For each of the reductions in the proofs of Theorems 7.3.11 and 7.3.18, show that there is a one-to-one correspondence between satisfying assignments for x and the appropriate type of solutions for $f(x)$.

7.3.29 It does not seem to be the case that $CNFSAT \leq_p 2CNFSAT$ (satisfiability for conjunctive normal form Boolean expressions with two literals per clause). Try to adapt the proof of Proposition 7.3.10 and consider what goes wrong. Then show that there is a polynomial time algorithm for deciding membership in $2CNFSAT$. (This is probably a very hard problem.)

7.3.30 The theory of addition was defined in the beginning of Section 6.4 (referring back to Section 4.1). In Section 6.5 this theory was shown to be decidable, while in Section 6.4 it was shown to be exponentially difficult. Consider the restriction of the theory of addition called the *existential theory of addition* which considers only sentences of the form $\exists x_1 \ldots \exists x_n F$ where F is a formula in the language of addition containing no quantifiers. Let ETA be the set of such sentences which are true in the interpretation of the natural numbers under addition.

(a) Show that ETA is $\mathcal{NP}$-hard by showing that $CNFSAT \leq_p ETA$.
(b) Show that ETA is $\mathcal{NP}$-complete by showing that $ETA \in \mathcal{NP}$.

Notice that this is the only $\mathcal{NP}$-complete problem in this section for which it is much easier to show that it is $\mathcal{NP}$-hard than to show that it is in $\mathcal{NP}$, in fact we know no easy proof that $ETA \in \mathcal{NP}$.

7.3.31 Consider ordinary regular expressions *without* the operation $*$. Show that the problem of testing whether such expressions represent the set of all strings of some given fixed length is $\mathcal{NP}$-complete.

7.3.32 The scheduling problem is an example of a problem which, even though it is $\mathcal{NP}$-hard, does have reasonably efficient *approximate* solutions. The set $T = \{t_1, t_2, \ldots, t_n\}$ of job times is easily sorted and so we may assume without loss of generality that $t_1 \geq t_2 \geq \cdots \geq t_n$. We denote by T_{min} the time required to complete the processing of all the n jobs for the optimal assignment of the n jobs to the k processors. Given that the times are already sorted with $t_i \geq t_{i+1}$, a *reasonable* algorithm, L, is one which simply assigns job t_1 to the first processor, and, after assigning job t_i, assigns job t_{i+1} to the first processor to become available. L is called the *largest-processing-time* algorithm. We denote by T_L the time required to complete the processing of all of the n jobs for the assignment given by L.

(a) Show that if in the optimal assignment at most two jobs are assigned to each processor, then $T_{min} = T_L$. (*Hint*: In this case it is very easy to see that *if* L also assigns at most two jobs to each processor, then $T_{min} = T_L$. You might try to prove the stronger result that if $n \leq 2k$ then $T_{min} = T_L$, but this is more than we need.)

We next state a series of results which show that L is always *nearly* optimal, specifically that $T_L \leq ((4k - 1)/3k)T_{min}$.

(b) First suppose that for some T, $T_L > ((4k - 1)/3k)T_{min}$. Show that in this case the inequality must hold for some T in which t_n is the last job to be completed in the schedule produced by L.

(c) Obviously $T_{min} > (1/k) \Sigma_{i=1,n} t_i$, and if we let τ_n be the time when the optimal algorithm schedules t_n to start, $\tau_n \leq (1/k) \Sigma_{i=1,n-1} t_i$. Why? Conclude that

$$\begin{aligned}
T_L/T_{min} &= (t_n + \tau_n)/T_{min} \\
&\leq t_n/T_{min} + (\Sigma_{i=1,n-1} t_i)/kT_{min} \\
&\leq (k - 1)t_n/kT_{min} + (\Sigma_{i=1,n} t_i)/kT_{min} \\
&\leq (k - 1)t_n/kT_{min} + 1
\end{aligned}$$

(d) Show that if we assume that $T_L > ((4k - 1)/3k)/T_{min}$, then elementary algebra yields $T_{min} < 3t_n$. But this implies that no processor is used more than twice in an optimal schedule, contradicting part (a).

(e) Show for some scheduling problems T that L does not produce the optimal schedule, and that there are at least some values of k for which the bound on T_L cannot be improved.

(f) Explain how to implement L in polynomial time.

References and Further Readings

For Chapter 1, more elementary treatments of many of the topics may be found in some of the following books. CLARK AND COWELL has a more detailed examination of programming techniques for Turing machines and other computational models. MINSKY has more on Turing machines and some useful intuitive discussions. YASUHARA treats Turing machines, recursive functions, and in particular, the elementary and primitive recursive functions. DAVIS is a collection of fundamental papers, including ones by Church, Gödel, Kleene, Post, and Turing.

For Chapter 2, the translation of Gödel's celebrated 1931 paper which appears in DAVIS gives an amazingly clear introduction of many basic ideas. The reader intrigued by Gödel's β function discussed in Section 2.3 is referred in addition to MENDELSON.

For Chapter 3, ROGERS' book provides a very thorough treatment of recursive function theory. ROGERS' paper lays the groundwork for a general theory of acceptable programming systems. POST, which also appears in DAVIS, sets the stage for a great deal of work in recursive function theory.

For Chapter 4, DAVIS contains the important early papers on the Gödel Incompleteness Theorem, including Gödel's original paper and Post's account of his own approach to the subject. Our proof of this theorem, based on the min-computable functions, is similar to one appearing in SCHOENFIELD. Elementary and more extensive treatments of mathematical logic may be found in ENDERTON, MENDELSON, and YASUHARA.

For Chapter 5, two fundamental papers are by BLUM. HARTMANIS AND HOPCROFT provides some good intuitive discussions, a different point of view, and some useful references.

Intrinsic difficulty results like those in Chapter 6 were first proved by Albert Meyer and his colleagues. MEYER includes a survey of some results and a partial bibliography. FISCHER AND RABIN is the basis for Section 6.4. ANDREWS treats elementary number theory, including Chebychev's Theorem used in the exercises of Section 6.4.

For Chapter 7, POST and ROGERS' book give motivation for the method of reducibilities and completeness. WINKLMANN gives an extensive treatment of the complexity of problems involving ALGOL-60 procedures, including the results in Section 7.2. COOK and KARP are basic papers on $\mathcal{N}\mathcal{P}$-complete problems and related ideas. GAREY AND JOHNSON gives a comprehensive treatment of $\mathcal{N}\mathcal{P}$-complete problems, including over 200 examples. BAASE includes a more elementary introduction to $\mathcal{N}\mathcal{P}$-completeness and some information about approximate solutions.

ANDREWS, G. E., *Number Theory*, Saunders, Philadelphia, 1971.

BAASE, S., *Computer Algorithms: Introduction to Design and Analysis*, Addison Wesley, Reading, Mass., 1978 (to appear).

BLUM, M., "A machine-independent theory of the complexity of the recursive functions," *Journal of the ACM*, Volume 14, Number 2, (1967), pp. 322–336.

BLUM, M., "On the size of machines," *Information and Control*, Volume 11, Number 3, (1967), pp. 257–265.

CLARK, K. L., AND COWELL, D. F., *Programs, Machines, and Computation: An Introduction to the Theory of Computing*, McGraw-Hill, New York, 1976.

COOK, S. A., "The complexity of theorem proving procedures," *Proceedings of the 3rd Annual ACM Symposium on the Theory of Computing*, 1971, pp. 151–158.

DAVIS, M. (ed.), *The Undecidable. Basic Papers on Undecidable Propositions, Unsolvable Problems, and Computable Functions*, Raven Press, New York, 1965.

ENDERTON, E., *A Mathematical Introduction to Logic*, Academic Press, New York, 1972.

FISCHER, M. J., AND RABIN, M. O., "Super-exponential complexity of Presburger arithmetic," *SIAM-AMS Proceedings*, Volume 7, 1974, pp. 27–41.

GAREY, M. R., AND JOHNSON, D. S., *Computers and Intractability: A Guide to the Theory of $\mathcal{N}\mathcal{P}$-Completeness*, W. H. Freeman, San Francisco, 1978 (to appear).

HARTMANIS, J., AND HOPCROFT, J. E., "An overview of the theory of computational complexity," *Journal of the ACM*, Volume 18, Number 3, (1971), pp. 444–475.

KARP, R. M., "Reducibility among combinatorial problems," in *Complexity of Computer Computations*, Miller and Thatcher (eds.), Plenum Press, New York, 1972, pp. 85–104.

MACHTEY, M., "Notes on crystalline glazing," *Ceramic Review*, 1978 (to appear).

MENDELSON, E., *Introduction to Mathematical Logic*, Van Nostrand, Princeton, N.J., 1964.

MEYER, A. R., "Weak Monadic Second Order Theory of Successor is not

Elementary-Recursive," in *Logic Colloquium: Symposium on Logic Held at Boston, 1972–1973*, Dold and Eckmann (eds.), Springer-Verlag Lecture Notes in Mathematics, Volume 453, 1975, pp. 132–154.

MINSKY, M. L., *Computation: Finite and Infinite Machines*, Prentice-Hall, Englewood Cliffs, N.J., 1967.

POST, E. L., "Recursively enumerable sets of positive integers and their decision problems," *Bulletin of the AMS*, Volume 50, (1944), pp. 284–316.

ROGERS, H. JR., "Gödel numberings of partial recursive functions," *Journal of Symbolic Logic*, Volume 23, Number 3, (158), pp. 331–341.

ROGERS, H., JR., *The Theory of Recursive Functions and Effective Computability*, McGraw-Hill, New York, 1967.

SCHOENFIELD, J. R., *Mathematical Logic*, Addison-Wesley, Reading, Mass., 1967.

WINKLMANN, K. A., *A Theoretical Study of some Aspects of Parameter Passing in ALGOL-60 and in Similar Programming Languages*, Ph.D. Thesis, Purdue University, 1977.

YASUHARA, A., *Recursive Function Theory and Logic*, Academic Press, New York, 1971.

Glossary of Unalphabetized Notations

NOTATION	DESCRIPTION	PAGE
Σ	alphabet	8
Σ^*	words over Σ	8
ϵ	empty word	8
$\lvert x \rvert$	length of word x	9
XY	set concatenation (for sets of words)	9
$X^{(n)}$	n-fold set concatenation	9
X^*	"star" operation	9
S^n	n-fold Cartesian product	9
a^n	string of n a's	9
$[x]^n$	n-fold concatenation of the word x	16
$-$	minus function for words	17
$\exists y/x$	bounded quantification for words	20
$\forall y/x$	bounded quantification for words	20
$\exists y \leq x$	bounded quantification for words	23
$\vert$	divides evenly	23
$<$	less than	23
$\leq$	less than or equal	23
$\exists y \leq f(x_1, \ldots, x_n)$	bounded quantification	24
$\leftarrow$	copy instruction for RAM	28
$\lvert \vec{x} \rvert$	length of the n-tuple of words $\vec{x}$	46, 169
∞	divergent computation	47
$\exists m \leq n$	bounded quantification for natural numbers	58
$\forall m \leq n$	bounded quantification for natural numbers	58

255

INDEX